Martians

By Steve Preston

3rd Edition

Table of Contents

Introduction

I have no idea why so many are covering up what is being found on Mars. It simply makes me mad. Mars has played a major part in our civilization and needs to be studied by our high school students so they can understand something about Earth's history. Not only was it a major influencer of our home, it most likely, was responsible for the massive hole in our planet we know as the Pacific Ocean and many of the mountain ranges that cover our world. Additionally, it is littered with the remains of what must have been a well-developed society and the details are not part of our student's curriculum. It is hard to image just how colonization began on this remote planet, but somehow people were there, substantial civilizations were there, and destruction was found to take away the lives of many people living on Mars. Even after the destruction, life remained and even flourished more underground than on the cold surface.

The atmosphere slowly was swept away to the levels we register today so people were forced underground, but there are signs that some life remained on the surface. We will investigate the first destruction of Mars, believed to have been about 400 thousand years ago; and its reemergence as a colonized world during our Pleistocene Age. During that fateful time, we can see the remains of a massive destruction as war eventually finished off many of the colonists leaving the obvious attempts to terraform Mars and the shocking remains of the war. While humans were lost, various animals and vegetation flourished in small pockets where the fleeting water could be found. At each place where water came close to the surface, trees of various types grew and can still be found as shown next.

Where water momentarily could find the surface, herd of animals could possibly roam and survive on the plants and the short supply of water. Some believe the following are herds of animals feeding on the sparse plant life on Mars. Who knows?

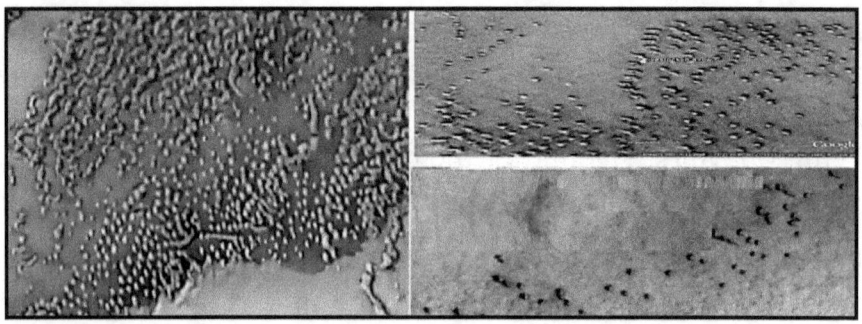

There may even be humans living on Mars. The first question might be how did they get there. We will look at that a little more later, but an effigy of a space ship found in Turkey might help here as taking trips outside of out atmosphere must have been fairly common before the

Bharata War of 3500BC. The Sumerians made the model below from something they saw. It is a vehicle similar to our space shuttle with 3 rocket engines out the back and an astronaut wearing full body suit. The Sumerians knew about these vehicles over 5 thousand years ago. I put flames and a face to the astronaut for effect, but there is little doubt that this vehicle was not for normal flying.

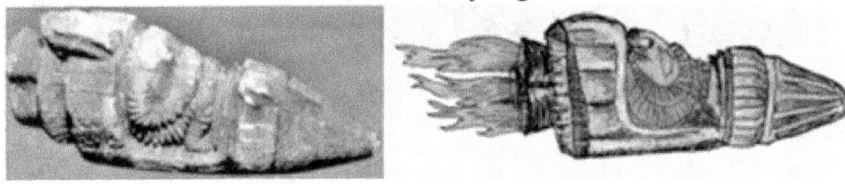

The reason these people colonized Mars possibly was for protection, or freedom, or excitement, or adventure, or whatever and they knew there was plenty of water and the vegetation make a reasonable amount of oxygen. From thousands of images showing civilization, some images show humans were momentarily found on the surface. Where you find human, you find mankind they always say, so I'm going to assume that is the most reasonable answer.

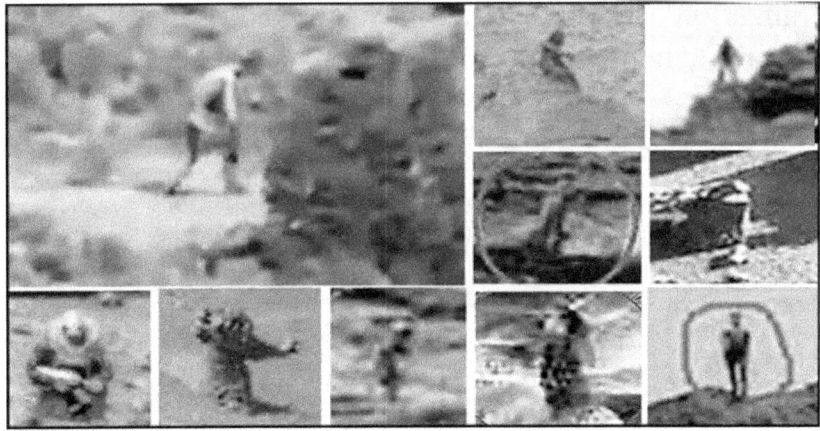

There is so much water that in the winter time, huge water vapor clouds are sometimes produced that are hundreds of miles long as shown next. The one recently captured by the orbiting satellite is over 900 miles long.

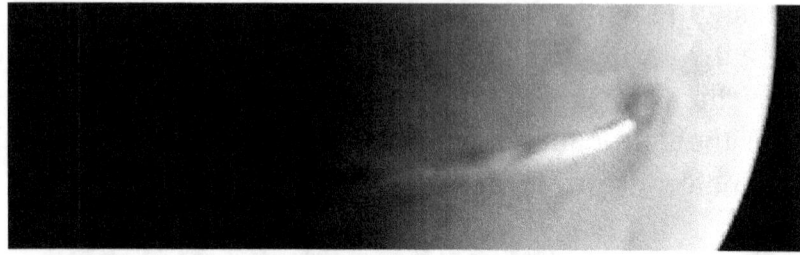

The remains of what appears to be a massive boat was even found.

Besides that, hundreds of buildings have been sighted and recorded. Here is one of the towns. Some of these images are really impressive, but somehow people still doubt of the colonization of Mars. The only questions that should be asked are, when was the colonization, and what happened?

The details of Martian civilization, culture, destruction, and current remnants of a once thriving community not only affects our understanding of where we came from, and how we got where we are today, but also; the large group of potential re-colonizers building for a near time blast off can gain solace in these details. If we could get just a little more detail in our images, we might even be able to read their ancient language as found on the inscribed block found on the Martian surface.

Before we get to the more recent details that give us a better understanding of Martian people who worked and lived on our nearest livable planet, let's start back a-ways as interest in Martians really started about 150 years ago.

1877 Schiaparelli [Canali]-It was then when Italian astronomer Giovanni Schiaparelli saw deep trenches meandering across the red planet's surface, which he called "canali." This meant deep grooves, but someone thought he was saying canals and it was believed if canals were on the planet, people were as well.

Lowell [Martian Life]-The American astronomer Percival Lowell continued Schiaparelli's work, but his idea was to popularize the idea that Mars held life. His enthusiastic interpretation of the canals as Martian constructions alienated his assistants and annoyed Schiaparelli himself.

Below are his detailed drawings of the deep canal like indications he "saw".

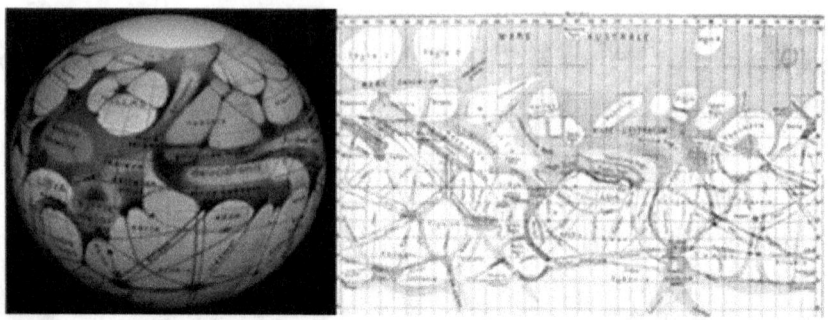

1898 [Article]-An 1898 article in *"The Atlantic Monthly"* noted that Mars might be in *"an advanced stage of evolution"* compared to Earth. We were now certain of the advanced Martians; we simply didn't know what their intensions were.

1906 [Book]-In 1906, Lowell published a popular book, *Mars and Its Canals*, proposing that the canals served to transport water from the poles to the planet's more arid central plains. They were proof of Martians being civilized and advanced.

1907 [A.R. Wallace]- In 1907, *Is Mars Habitable?* by A.R. Wallace directly critiqued Lowell's book, and presented Natural reasons for the apparent indentations.

1938- [Orson Wells]- To keep people from hysteria, Orson Welles developed War *of the Worlds*. Wait a minute! He did the opposite thing. According to his broadcast, the canals had been signs of a more developed, and hostile group of metallic Martians. Mars had become a dry, red, craggy landscape populated only by nasty Marians bent on our destruction.

Water on Mars-The concerns over Martians began to fade and scientists found no water to support life, but recent news of liquid salt water on the planet's surface has people wondering again about the Martians and what they might want with us.

Hysteria over Mars has reached a breaking point over the past couple of years as NASA's Curiosity rover beamed back breathtaking photos to Earth. Another rover will be sent up in 2020 to find more about life on Mars. To really understand Mars and the potential for Martians, we need to look at a time before people and before dinosaurs. From that time, we need to expand to a time when people gained great advances in technology which eventually led to war. With destruction everywhere, some people, apparently, went to Mars. We will trace the possibility of the evidence, stories, and details that may help solve the history of Mars and Martians.

Unfortunately, the Martian atmosphere was losing more oxygen than could ever be replenished and over a long time, the planet all but died. Today there are signs of terraforming and some level of civilization still in place and there is the Rainbow Project and the Pegasus Project. With such a large number of people claiming to be part of this mysterious event, we cannot ignore the probability of life on Mars.

Timeline

So, you won't have to spend too much time in disbelief, let me provide you with a general timeline. It's ok to doubt these events, but if you keep an open mind, I think you will understand more about where you live and about our 2nd closest planet, Mars. This timeline covers events on earth, Venus and Mars because they are all connected in some way. Venus, of course abruptly halted any influence it had when it caught on fire about 11 thousand years ago, but it used to be a pretty good planet from what ancient texts tell us. Mars also had its day, but all things come to an end. The signs are still visible and the evidence is still around. Before I get started, let me say that the timing you had once believed can no longer be supported after the finding that nuclear decay is NOT constant at all. After retiming the events on earth, we find that the dinosaurs, generally were destroyed at the end of the Cretaceous Period, but it was only about 120 thousand years ago. Don't worry about that right now, but keep it in the back of your mind.

Long Ago-The planet Mars was formed along with Earth and the others. The planets had more elliptical paths than they do today [More like that of Neptune today]. With the odd revolutions, there came problems.

Beginning of the Permian Age [about 600 thousand years ago]-This is believed to be the first close encounter caused by the less than stable orbits of Mars and Earth. This produced the American Ridge Mountain ranges. On earth 90% of all life was destroyed. Mars could not have fared well.

The End of the Permian Age [About 440 thousand years ago]-Mars and Earth almost collide again and the Pacific Ocean was formed. Half of Mars split away. Particles became the Asteroids. This event also established the Himalayan Ridge Mountain range. Both planets began rotating much faster than they originally had rotated. The faster spin made everything on both planets lighter, but it also made the atmospheres of both planets begin to slowly dissipate into space, just like Saturn's atmosphere is doing today. The Earth had a huge gash in it called the Pacific Ocean and Mars had been torn in half. Anyone or anything living in the Pacific region of Earth or the entire Northern hemisphere of Mars lost their lives. In Antarctica, we took Ice core samples and found that before this time the earth was much different which goes along with about ¼ of the planet being yanked into space and the massive continent I call Prestonia disappeared forever as the Asteroid belt came into existence.

Triassic Age [around 380 thousand years ago]-Besides animals reappearing, evidence shows that mankind came about around this time. These weren't ape-men, but regular old stand up straight, using a calculator type people. One shoeprint fossil was found with 2 dead trilobites underneath it. The human wearing the shoe walked on these things and killed them as trilobites became extinct during this time.

Men walking on trilobites wasn't the reason for their destruction.

Jurassic Age [around 280 thousand years ago] --Huge cities were established on Earth by this time. People used electricity, had modern construction methods, understood genetics, they developed flying machines.

Cretaceous Age [around 180 thousand years ago] --We believe Mars was colonized by this time. Unfortunately, like the Earth, every year the atmosphere got lighter and lighter as the planet was spinning too quickly to support the atmosphere now that half of it was gone. On Earth man had developed useable nuclear power. That would be a problem.

Cretaceous Extinction [around 120 thousand years ago] --We read about massive wars on Earth. 85% of all life became extinct and huge meteors hit its surface. Mars would have had a horrible time as well as they depended on the Earth for supplies as their living conditions worsened. On earth, everything was gone. On Mars the southern portion of the planet remained pretty much intact.

Tertiary Period [120 to 40 thousand years ago] ---Earth was saved because its rotation slowed. The rotation of Mars, on the other hand, stayed faster and more of its atmosphere left every year. After the war, we can believe many migrated to Venus and Mars to build a new life. Many stayed on the earth as well.

Pleistocene Period [40 to 10 thousand years ago] ---A second massive war erupted. This would not be isolated to Earth. We are told many troops were stationed on Venus. We can assume the distance to Mars isolated it, but by this time there was almost no breathable oxygen and people either had all come home or were trying to survive in underground dwelling. The war is failure and the moon of Venus was destroyed somehow. Pieces of the Venusian moon shattered everywhere. Hundreds of thousands hit the earth and ushered in the Holocene epoch. About 1/3 of the population of the Earth was killed and the Carolina Bay craters are still found all over the east coast of the United States to mark the fateful day. Almost everyone that lived on the planet Venus was destroyed. One group called the Dropa, evidently escaped and fled to Earth just before the explosion. They landed to the hills of China, but that is another story. The picture following shows the way Venus is today. No one is alive and no cities survived.

Pleistocene Extinction [around 10 thousand years ago] -- A third major war was initiated on Earth or the war continued. It was more devastating that the one that ended a few hundred years earlier. The nuclear fallout from the war was everywhere. We are told massive mutation in human DNA changed people. We are also told at least 1/3 of the entire population of the world was killed. We know that the Earth shifted about 30 degrees on its axis and Mammoths grazing in a field were thrust into the Artic and were frozen so fast, they still had flowers in their mouths as found in Siberia.

Bharata War [3300 to 3100 BC]-After the Earth settled almost no one survived. We are told about 10 percent of the Anak people survived, a number of hybrids sometimes called Gentiles, and only 8 of the pure Cro-Magnon people led by a man named Noah. It would be another 4 thousand years before great civilizations would be restored. Like the other times, greed, power hunger, and war ensued. Nuclear war and biologic destruction killed about 1/3 of the people again but this time something happened to the brain as massive mutation again is recorded in our DNA 55 hundred years ago. On Mars, we can believe most or all or those trying to survive had lost the race.

Holocene Events on Mars-Mars had become a wasteland during the Pleistocene war, but some remained. During this time, Mars appears to have been change as the signs of terraforming are still visible on the planet today.

Possible Research-By the 1940s, man had begun to understand the concepts of Nikoli Tesla and time travel, of a sort, was reportedly, established during the "Rainbow Project" and its subsequent testing event known as the "Pegasus Project". The reason this is of interest to this book is that we are told, some of the Project-subjects were "assigned" to work on the planet Mars. I will provide an overview of this work at the end of the book for reference.

Before you can really understand Martians, you need to first keep an open mind and also you need to be retaught about the Earth's timing.

A New Timing

You were told nuclear decay as a form of "accurate dating. For some time now, ALL have known the huge issues in this previously established timing baseline. You have been told dinosaurs died 65 million years ago and the beginning of the Mesozoic Era was 300 Million years ago. You were told, over and over and over again. They even proved to you that was truth by telling you lead, potassium, and even carbon isotopes decayed at a set rate; just see how much of an early isotope is left and read the date. Besides, the dinosaurs are buried underground and turned into stone so there had to be a long time for that to happen. Besides finding dozens and dozens of dinosaur-remains that are not even fossilized and the whole timing accuracy thing was all a lie, our children have been given all they need to understand the world.

Ancient Earth

While the earth is ancient, it is definitely not as old as has been told to you. Many of geologists today still tell you that radiometric dating has narrowed the age of Earth to about 4.5 billion years, give or take a couple of percent. We now know that the dating method is inaccurate and scientists not pursuing that vain truth I talked about earlier, are refining the timing more and more each day. The Earth and everything in it are much younger and so are the characteristic stabilities of the planets in our Solar System.

Researchers at Purdue and Stanford have found evidence that **radioactive decay rates are not constant at all.** <u>On December 13, 2006</u>, a magnificent solar flare flung radiation and solar particles toward Earth. Measuring the decay rate of manganese-54 during the flare proved to be very interesting as the decay rate dropped during the time of the radiation fallout. It was determined that solar neutrinos zipped through space and affected Mn-54's decay rates used in the experiment. Just think about this. They were testing a single solar flare event and the change was significant. The sun has these things all the time. It was also found that the decay rates of silicon-32 and radium-226 showed seasonal variation, according to data collected at Brookhaven National Laboratory on Long Island and the Federal Physical and Technical Institute in Germany. This error was just the material sitting there with almost no outside interference. Wood buried in igneous rock, at Queensland Australia, has been dated to <u>40 thousand years</u>, while the basalt around it dated to <u>45 million years</u>. Excess argon-36 was found in three out of 26 lava flows in recent times. So, <u>Argon/argon testing would show a much older date that actually was "KNOWN"</u>. This is believed to be because there was too much of the argon-36 in the first place. In the Grand Canyon, lava flow testing showed lower levels of lava were younger than the top layers. At different volcano sites, that had eruption in 1949, 1954 and 1975. The same thing was noted. Geochron Laboratories of Cambridge, Massachusetts dated these samples. Even though the <u>oldest of these samples are just over sixty-years old</u>, the lab tests provided ages that <u>ranged from 270,000 years to 3.5 million years old</u>. Additionally, we go to Mt. St. Helens and its eruptions in the 1980's. Samples there gave old ages in the range of <u>300,000 to 2.7 million</u> years. Hopefully, you are beginning to see that we know less about how old we are than you believed before reading this. If neutrinos from a single solar flare can make things look older, what if the entire Earth was closer to the sun? I know that sounds odd, so just keep it in the back of your mind right now as we try to find some standard for dating.

Today we know that the nuclear decay dating of things including Electron Spin Dating and Uranium Dating, Thorium Protactinium Dating, Oxygen Sediment Dating, Lead-lead-lead Dating, and Argon Dating, which we originally used to date the ages of the Earth, are terribly flawed. The old standard carbon 14 dating also seemed in jeopardy. Dating beyond about 30 thousand years was <u>much younger</u> than tested. If there had been nuclear events [bombs or even volcanic eruptions] the apparent timing was changed drastically. Other methods had to be employed to determine how everything should be timed, but classroom information was not changed. That would confuse the students. I'm going to prove to you how you have been lied to. This will give you a better understanding of the lengths some will go to when they believe something, no matter what the evidence shows.

Standard Geological Timeline

Era/Period/Epoch	Time (M yrs. ago)	Time (T yrs. ago)
Archaeozoic Period	5000-1500	50,000-3000
Proterozoic Period	1500-545	3000-1000
Cambrian period	550-500	1000-900
Ordovician period	500-440	900-800
Silurian period	440-410	800-700
Devonian period	410-365	700-600
Carboniferous	365-300	600-500
Permian period	300-250	500-400
Triassic period	250-212	400-300
Jurassic period	212-145	300-200
Cretaceous period	145-65	200-100
Tertiary period	65-1.8	100-40
Pleistocene period	1.8-0.01	40-10
Holocene period	0.01-0	10-00

The middle listing of dates is the "STANDARD" that had been presented in our classrooms [in millions of years], while the last column shows a somewhat closer, more accurate time line that has been verified by non-nuclear decay methods [in the order of hundreds of thousands of years.]. Even with the mountain of evidence showing how nuclear decay cannot be used, the middle timing is still heralded as the master in many schools and books being used to teach our children without basis. I know it is difficult to believe historians, scientists and teachers would keep these things from you, like how does greenhouse gas affect our planet, so let me tell you a little more as I try to open up your mind to possibilities that you have been lied to.

Stratigraphic Positioning

Besides Nuclear decay, the main way scientists used to determine "age" was by Stratigraphic Positioning. This is the determination of age by position, depth, and material consistency. MANY TIMES, this is the only method for cross comparison that was thought to be reasonable for confirmation of Radioactive decay. Scientists simply determine the depth of objects, or features near the object, or number of lava flows, or similar geologic characteristics and use the depth as a time gage. This type of comparison may not have a very high level of accuracy, but seeing things in different layers seem to show when something died. If something is lower, it is older and newer is newer. Added to this method is something called the K-T boundary, where iridium chalk was deposited from an ancient meteor that struck the Yucatan around the time the dinosaurs died. Scientists have been using this for a long time when, all of a sudden, there were trees found that were going the wrong way. The next set of pictures shows some of the unfortunate trees that must have died repeatedly to be deposited perpendicular to all of the stratigraphic lines.

Some try to state the trees simply fossilized while standing for MILLIONS of years as the ground built up around them. [20 points on the BUNK meter!]

Distance to the Sun

If neutrinos from a single solar flare can make things look vastly older, what if the entire Earth was closer to the sun a few hundred thousand years ago? I know that sounds odd, so just keep it in the back of your mind right now. Right now, I'm going to provide you with a more logical way the Pacific Ocean was made at the beginning of the Triassic Period as our planet revolution around the sun was not stable. That is where Ice samples come in.

Ice Core Dating

Although the task is tedious, ice can be examined just like tree rings. Each summer ice changes its consistency. H_2O (16) is more concentrated in the summer while H_2O (18) is more concentrated in the winter. This gives us indication to the level of CO_2 which in turn allows us to understand something about the temperature levels. As the yearly cycle has freezing and thawing, ice consistency varies each day, seasonally, and yearly, depending on Earth axis and other critical elements. Anyway, scientists around the world started boring holes in ice. The most coring is done in Greenland and Antarctica. A sample is shown below.

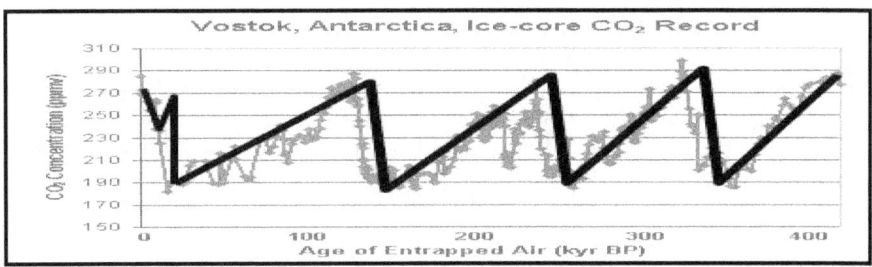

If you look closely you can see that about every 120 thousand years there is a major change in the environment. This was found at both Antarctican and Greenland Ice cores and the dating is by seasonal changes rather than nuclear decay. Bah humb! You say! Well, what if we see confirmation?

Hawaii Hotspot Track Dating

Hawaii is not a tiny group of islands, but instead is an indicator of where the Earth magma has a hotspot. As the crust moves differently than the stuff below, the hotspot relative to the crust moves and each time the hotspot burns through another piece of crust, a volcano erupts which seals off the area after a time and an island is made for a few thousand years. This travelling hotspot known as Hawaii is show next. The descriptions provided shows what was happening along the way. Because the hotspot moves perpendicular to the axis of the Earth, we also know how the earth was spinning as shown by the lines in the first graphic below, but the actual timing is not described here. I placed some general times in the second graphic, but let's see if they make sense.

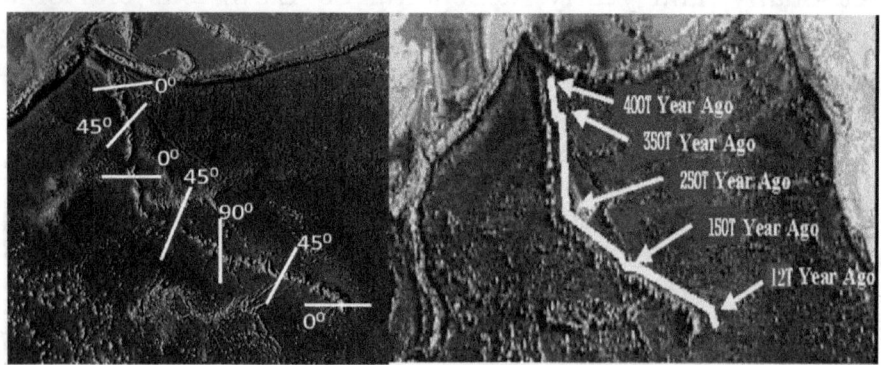

Let's compare the Earth shifts with the Ice core data. Man-oh-man; it seems they match. I think you still believe in nuclear decay so we will look farther.

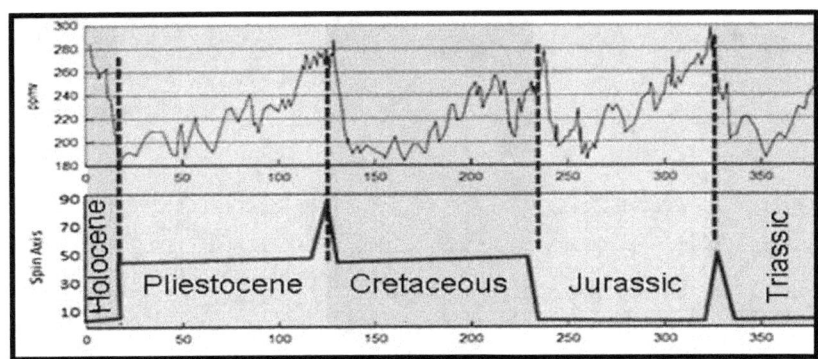

The Atlantic Ocean is getting wider about an inch a year, averaged worldwide. While the building of the great mountains has little to do with the normal tectonic plate "drift" We can pretty accurately measure the widening ocean in various ways including measuring distances between matched magnetic landmarks on either side of a widening gap on the ocean floor. The old theory indicated that 180 million years ago the continent Pangea began splitting apart and has been drifting ever since. In so doing, the landmasses of the Western and Eastern hemispheres separated and opened the Atlantic Ocean basin today.

Plate Tectonics

Plate tectonics tells us the outer hard crust of Earth consists actually of a dozen or so distinct, hard plates that drift individually on hot, deformable rock. An unequal distribution of heat within Earth moves the plates. The boundary between the plates forming the Atlantic Ocean is smack down the middle along the Mid-Atlantic Ridge, shown as the hashed line in the figure below. The ridge is where we must look to find a widening gap, which accounts for the widening ocean.

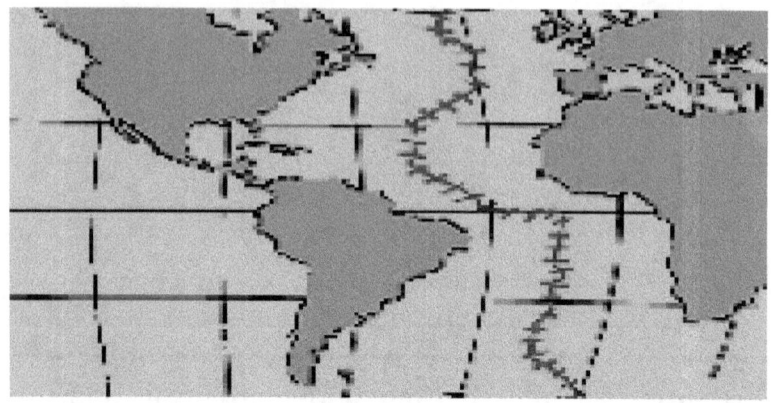

That is where we measure the rate of separation. Where the plates separate, white-hot soft mantle oozes up from great depths within the Earth to fill the gap. The molten rock cools slowly into new slivers of sea floor. This happened over and over again through the eons. That's how the Atlantic Ocean widened-by a spreading sea floor. Iron-rich rock has a peculiar property; heat it above its curie point of 580 degrees Centigrade and it loses its magnetism. When it cools the rock gets re-magnetized in the direction of the existing Earth's magnetic field. So, it's a magnet with the poles aligning with the poles of the Earth at the time of the cooling. The neat thing about this is: the magnetic field of the rock, once cooled, stays frozen in this orientation. It becomes a record of the Earth's field at the time of its cooling. The first graph below shows how the magnetic field has changed over time. Certainly, we cannot get an actual time, but a relative timing is very good. What if I told you this matched up exactly with the Ice Core and hotspot data?

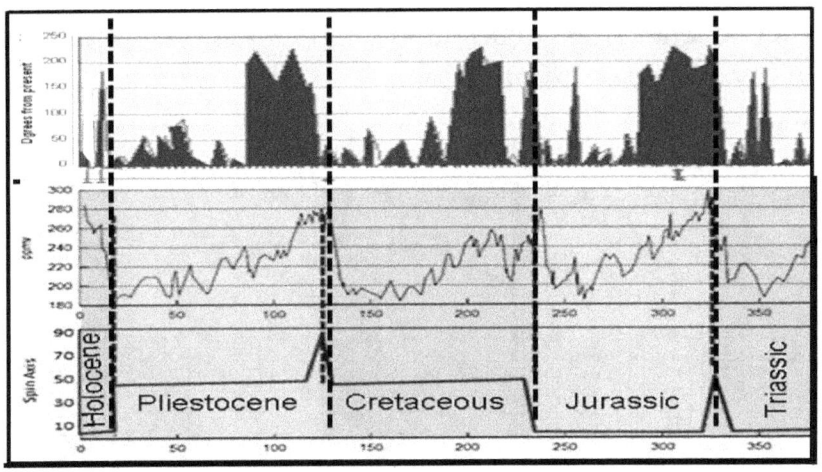

Marine Isotope Stage [MIS] Dating

Some people may still be reluctant to give up what the schools have been preaching so very long, so I thought I would bring out one last attempt at presenting sanity. Large numbers of scientists around the globe are doing Marine Isotope Stage timing by digging in dirt. It seems looking at the levels of Oxygen 18 shows how hot or cold a point is in time while checking relative Oxygen 18 isotopes in Calcite [which just happens to be the main ingredient in seashells], one can tell just how many of the things were here during each period. Checking around the globe has given us a good map about climate and number of seashells, which correlates to number of animals in general so it is easy to see where extinction periods are. Guess where they line up? Time's up! They are an almost exact match as shown below. MIS levels are shown next above the ice core samples, the hotspot data and the magnetic field shift data Massive drops in O_{18} mean massive drops in sea shells and all other life. Notice there is no extinction period between the Tertiary and Pleistocene Ages marked by Cro-Magnon appearing. Please say you see a comparison.

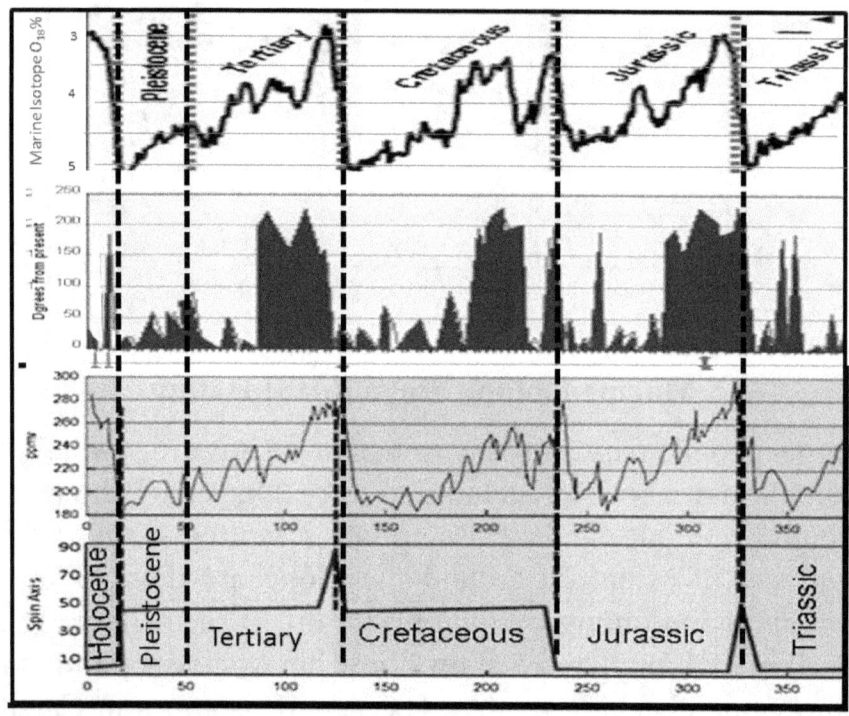

I know these dates are not what you were told and I know the idea of using nuclear decay to date things has made a comfortable geologic record, but the pieces don't fit. The mutation of mankind doesn't fit and new discoveries of non-fossilized dinosaurs don't fit. Another thing that is odd is that there should be an easy distinction in the geologic record to show when 1/3 of the planet was ripped away.

The Big Change

I'll bet you are wondering how in the world all these similar timing components could have been known and you were never told!!! If you are beginning to see how scientists, historians and professors will hold to an old "truth" even when it is found to be in error as they built their entire understanding of the universe on these now destroyed truths. Now let me continue this same tactic to help you see the probability of Prestonia being sucked into outer space by Mars. Scientists have gone even deeper into the Ice to find out about more ancient times. They show something very interesting as described in the following graphic. The chart below shows a cyclic changing in the earth's temperature and Carbon dioxide levels over the last 800 thousand years, but notice something STRANGE! What we see is that after 400 thousand years ago, there are very distinct and abrupt thermal changes every 120 thousand years associated with massive extinction periods. The cyclic nature continues before that time, but the events are greatly softened showing the characteristics of the Earth were vastly different before that fateful time. Possibly, this would be something about the larger planet and a smaller portion of the planet core being shifted as the earth spins. Can you see the remains of Prestonia going into space and the Planet Mars being ripped in half?

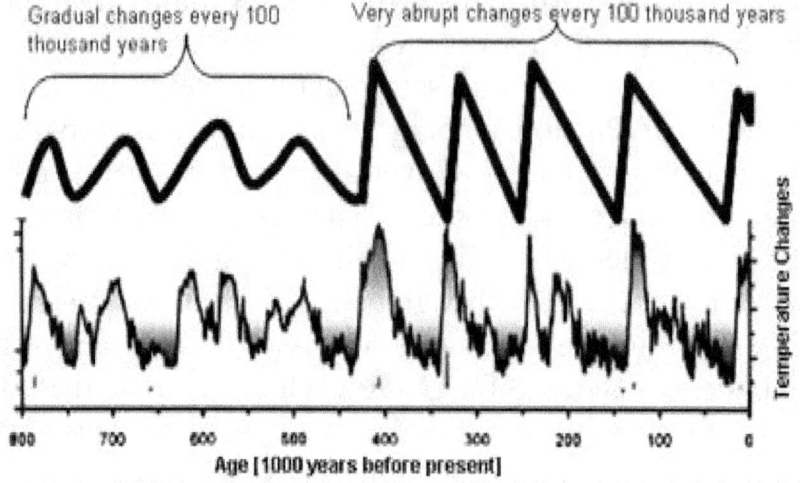

Mars Makes Mountains

Let's get back to Mars and back up in time a little; as this book isn't about the end of the dinosaurs, the war in heaven, the birth of Cro-Magnon, Venus and Venusians, or other wars. Here is something we can be certain of, Venus, Mars, and our moon must have been populated at one time, but it is the destruction of Mars that has been determined to be one of the major causes that Earth is livable today. This is going to require a little explanation. Let me go through a short timeline of the Martian encounters throughout history. Let's first start with The Rockies and the Andes Mountains again.

Andes Were Made

500 thousand years ago the earth had its worse extinction period of animals ever. What in the world caused it? Well-- The planetary orbits weren't nearly as stable as they are today and models of near collisions of Mars and the Earth indicate that the 2 planets came so close together, parts of both planets _almost_ ripped away from the surfaces. On the earth, the western coastlines of the Americas were along the closest path of the planets and the massive range of mountains along that region were PULLED into place from the Martian gravity. The mountain range goes more than 50% around the earth and there would be no "direction" of plate movement that could have sustained the mountain

pushup theory, so don't even think what they told you has any truth. On Mars, the same would have happened as the gravity of Earth would have pulled on its surface. Oddly, there is no clear mountain range on Mars, so this sounds like bunk. Mars is smaller than Earth so whatever happened on the Earth would have been far worse on the planet Mars. Don't worry about that right now, it will make sense shortly. [Trust me!]

You would think there would be some evidence of mountains rising high in the air during this time period and there is. Before the first "uplift" from the Mars encounter, the western side of what is now South America was underwater rather than being mountainous. As the mountains were pulled up, huge prehistoric clams the size of a man were pulled up with the rest of the mountain. Today, evidence can be readily seen. Below are piles of huge clams on top of the Andes Mountains.

Mars Destroys Prestonia

We are now about 400 thousand years ago the positions of Mars, Venus, and Earth were still not where they are today; in fact, most of the revolutions of planets surrounding our sun were oval which caused issue for a time. Gravitational pulls from both of our nearby planets have caused great damage at one time or another and this would be one of them. If you have ever wondered where the massive mountain ranges of the Himalayas" came from this is the culprit. You were told the plates of the earth shifted to push the mountains 5 miles into the air, but the reasoning is faulty. While I don't want to get into a science study of the very real repositioning of plates of the Earth's mantle, let me just say that those telling you that plates were pushed in opposite directions simultaneously to build these extremely long ranges of the Americas and were pushing the crust up 5 miles into the air where the Himalayan Mountains exist were not telling you the truth. Instead these two ranges, at least, were pulled up from opposing gravitational forces of a nearby massive object. It is believed that about 400 thousand years ago, Mars orbit brought it fairly close to the earth for a short time. The gravitational pulls of both planets tugged on the planet exterior surfaces. Along the equatorial regions of both planets massive mountain ranges pull up and perforated the exposed land. Then came calamity. About half of the

Martian surface was yanked away and well over 1/4 of the Earth's surface pulled away as well. A portion got caught up as a satellite we call the moon, but most chunks mixed together and were swept out in space until the sun's gravitational pull began the orbit in the area known as the Asteroid belt. Please see the graphic I made next. The chunks that left of Earth, had once been the **Super Continent called "Prestonia" by some.** [I'm embarrassed to say I'm the only one calling this supercontinent by that name, but many ignore its existence altogether, so I had to call it something.] The remaining land "Pangea", almost immediately began to split apart to fill in the hole. This is the type of catastrophe that could cause our planet to lose its hemostasis and blast the earth temperature into a spiral that would destroy mankind. No hydrocarbons were used and the greenhouse effect was not able to convert the Earth into a ball of fire like Venus would thousands of years later, but we know that most of the animals around the world died as the end of the Permian Extinction left little in the way of life. ON Mars, it was even worse.

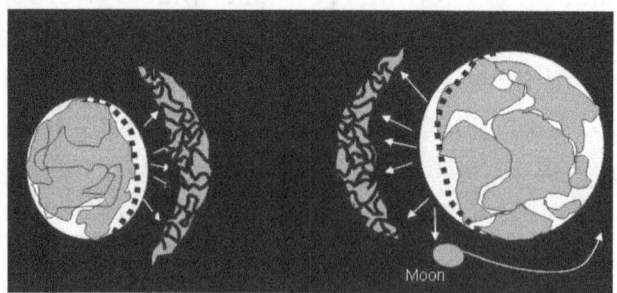

Proof of Prestonia

If you were one of those who thought that the super continent of Pangea was on one side of the Earth and nothing was on the other before this major event in Earth's temperature, weather, and life model was modified, you would have a very wobbly earth. Another super continent HAD to be opposite Pangea. While it is difficult to determine when all this happened, we can be pretty sure massive extinctions occurred. One way to time the massive reduction in mass is to look at Ice Core Samples and see when there was a change. Review the previous chart and see that the dynamics of the Earth changed a little over 400 thousand years ago. The image below shows where this mountain range [supposedly pushed up by plates getting on top of other plates.] Now just for kicks, make this mountain range MUCH bigger as it would have been made before Pangea began separating and making the Atlantic Ocean.

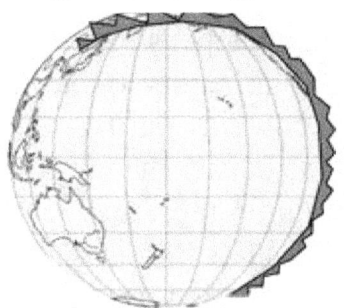

I Need to Bring up Something Important

After the Mars incident happened, the Earth and mars would have done something somewhat strange. Like a figure skater doing a spin, as the Earth and Mars got smaller, it began spinning on its axis much faster. This faster spin, made the gravity of the Earth lower so animals had to compensate.......... They got bigger. During the Mesozoic period following the creation of the Pacific Ocean, Massive dinosaurs erupted seemingly overnight. I know "scientists" have told you the Tyrannosaurus Rex could not run for fear of falling on his massive head, the Diplodocus and Pachycephalosaurs could not lift their 50 foot necks as blood could not flow, and the Pterodactyl could not fly as it was simply too heavy, but this also is a lie as all would have died if they could not run, fly and raise their heads to eat. General images of these anomalies are shown below.

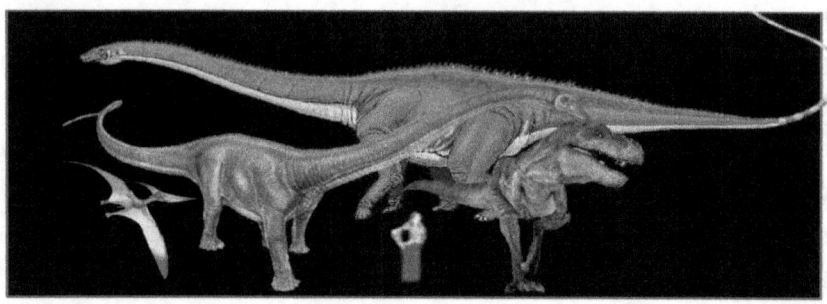

The other thing that happens when your planet spins too quickly is your oxygen begins to leave. Studies have shown that the oxygen content during the beginning of the Mesozoic period was denser than it is today. Dinosaurs with small hearts could pump more oxygen to cell because there was more OF IT! On Mars the same thing would be noted as we would find a substantial amount of water and air and atmosphere on Mars during the Mesozoic, but soon most would abandon the planet just like the atmosphere of Saturn is doing today as we watch.

Unstable Orbits

We now know that there is a 100-thousand-year cycle that seems to be self-generated by the Earth itself, sometimes the changing characteristics that we can use as time-marks. Every 100 thousand years or so, the Earth gets terrible cold, which sometimes shifts the rotational axis and causes extinctions and or problems. Mathematical models describe this change as having extraterrestrial connections. I know that sounds like little green men, but that isn't what I'm talking about. I'm talking about planets getting to close to one another. The following image shows the first of three of the major ones believed to have been associated with major events most likely involving Mars. To the right shows a second one that happened just before the end of the Pleistocene. While many have pushed these events back hundreds of millions of years ago, from the timing corrections presented in this work, we can assume that the stability of the solar system is only recently been "standardized" into almost circular orbits around the Sun. It is the variable intensity of the sun that has caused much of the confusion in timing, so let's peel back the history just a little.

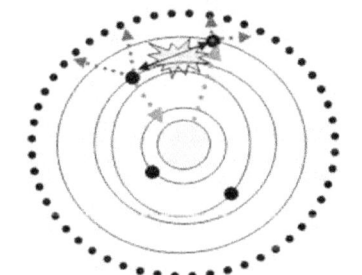 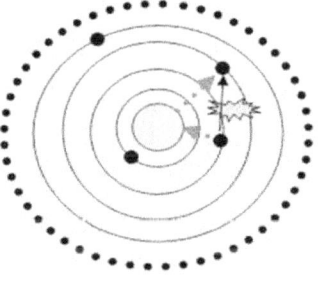

400 thousand years ago Mars almost hits earth-both planets have massive damage- Mars orbit pushed out, Earth orbit come closer to sun- Moon and Asteroids formed

12,000 YEARS AGO- close encounter Venus and Earth. Venus moved closer to sun, Earth farther away. Venus moon destroyed

I know all this stuff sounds foreign to many who are reading and I will briefly provide some of the details. There are many books on the things I have addressed and I have written books on these subjects as well. Before we go on, I have to bring up something else you should have been told about in school, but somehow, they forgot to tell you. We know people lived with the dinosaurs. <u>There is NO DOUBT</u> as thousands of pieces of physical evidence, historical reference, and religious documents **all** tell us exactly the same thing. Like the animals; people living during this time grew HUGE. Some called them *Titans*, the Book of Genesis called them *"Giants of Old"* [Genesis 6]. More oxygen, less weight both caused everything on Mars and the Earth to be larger. Not being able to see Mars well has limited our knowledge about and home-grown life, but we can imagine that the ancient people began colonizing the nearby planets very early on. Before we get to that let's look at some evidence.

Even with the above evidence many will say—Bah Humbug—to this whole concept. The near collision of Mars and Earth must be false because physical law requires that the smaller planet would have sustained the greatest damage. As we have already explained, THEY ARE RIGHT; at least in their theory. Let's look at the remains of Mars for a minute. I already mentioned that only half of the planet has significant amounts of cratering, but what I possibly didn't emphasize is the fact that while the earth split open during the flyby, Mars got the worst of it. It essentially split in two. That fact is so obvious, it is almost comical.

Thin Crust

Just like the Pacific Ocean on earth with almost no crustal mass remaining after the split, over the ENTIRE northern hemisphere of Mars, the crust is rarely more than a few kilometers thick and the sparsely cratered surface is suggestive of a relatively new surface. Like the remaining portion of the earth, not including the Pacific Ocean, the southern hemisphere of Mars has a strikingly thick crust, which exceeds 20 kilometers in places, and a much more heavily cratered surface. It is in this hemisphere that we find nearly all the major impact basins such as Hellas, Isidis and Argyre with crater basins well over 1 thousand Kilometers in diameter. These huge holes were probably made by some of the large chunks of earth that left during the explosion. As the graphic shows, after its last near collision with earth, Mars also became a new planet, much smaller than it had once been.

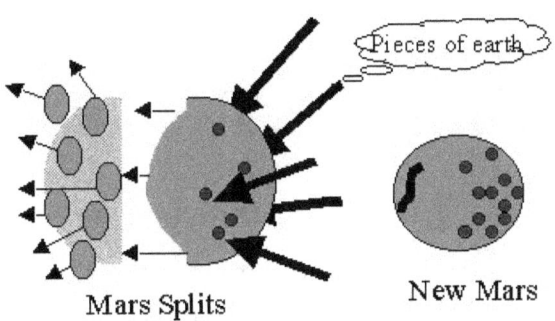

Mars Splits   New Mars

Evidence of a Split Planet

This odd cratering isn't most obvious evidence. The next image shows the proof from NASA. I know all of this sounds bizarre, so let's look at a topographical image of Mars again. Please note that the northern hemisphere is not only smooth, but it also is sunken in much worse than our Pacific Ocean. It has a mean surface height 6 thousand meters lower than the mean of the southern hemisphere. Where do you suppose the northern half of the Planet went to?

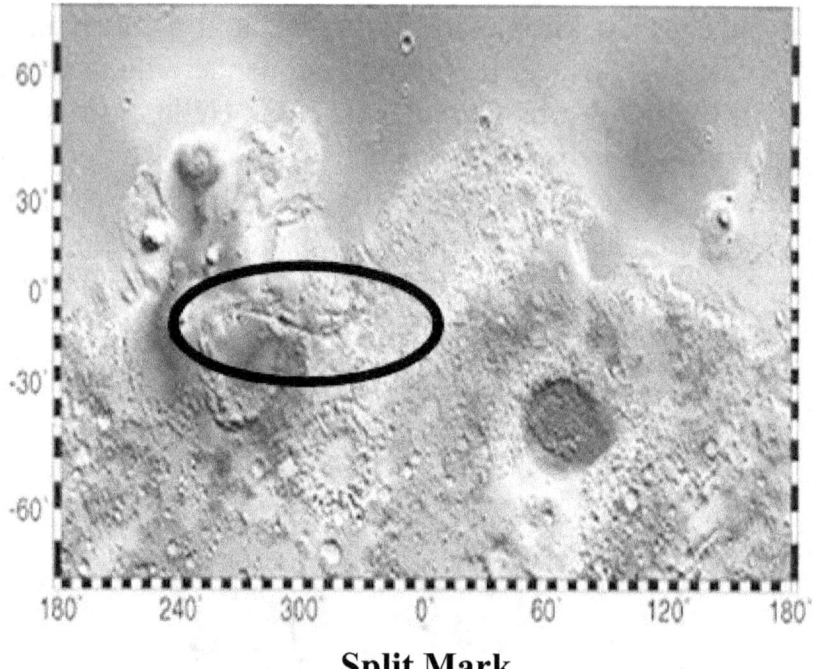

Split Mark

Also look at the dividing line between the hemispheres. Just like the huge slash along the edge the Pacific known as Mariana's Trench, the northern hemisphere of Mars, is marked by one huge gash called the Valles Marineris. [Circled in the preceding picture] Although it is only 7 kilometers deep, it is up to 200 kilometers wide in places and has a total length of 4,500 kilometers. [That's as long as the United States.] It is almost like Mars was split apart at one time and one of the marks of the split is this huge gash just like the Mariana's Trench in the Pacific.

Healing Mars

Mars, like the earth, is slowly trying to heal itself from this ancient event. In another 500 million years, the entire surface will probably be about the same height and here on earth, the Pacific Ocean will be about the size of the

Atlantic. The diagram below shows the general topographical distinction on Mars at the present time. The green area is the high area and it is slowly resurfacing the planet. As it does, a hole is opening at the South Pole.

Some Say Nothing Happened

Many scientists ignore the probability that Mars was split in to. There is a small possibly Mars has never split, but the lack of craters on 50% of the planet, the heavy cratering on the remaining half, the low surface area on the crater naked side, the huge fissure that is still there, and just about everything else tells a different story.

Review

Five hundred thousand years ago the orbits of the planets weren't circular and on rare occasions, the planets came close to each other. The sun was more intense during this time period and pushed out many neutrinos into the earth's atmosphere making more and more nuclear isotopes rather than allowing them to "decay" as many had previously believed. Hundreds of thousands of years appeared to be hundreds of millions of years because of this massive illusion. The earth during this time was filled with 2 great continents. The first will be known as Pangea, but the second one, on the opposite side of the earth had no name, so I named it Prestonia. Without the second continent, counter balance, the earth would not have been stable and would have wobbled violently. With all the close collisions going on, the Pacific Ocean was made. Mars got too close and pulled Prestonia right out of the earth and then Venus had its turn later, but we are focusing on Mars here so we can check out the Martians.

Timing The Last Encounter

Scientists are continuously measuring the speed of this continental separation. The Atlantic Ocean is the effect of the separation and its width corresponds to the Pacific Ocean creation. Here are some of the things we know:

1. There is almost NO crust under the Pacific Ocean, so we know the crust is missing from the planet.

2. We know the massive continent of Pangea could never have been created without an equally large continent on the opposing side of the Earth or the earth would be unbalanced. I call this missing continent Prestonia, but you can call it something else if you like.

3. We know that something horrible happened to end what is called the Permian period, when 90 to 95 percent of the animals died.

4. We also know that something very peculiar happened after this horrible time. Something happened to the animals and they all began to get huge. This oddness suggests that the earth began to spin much faster than it previously had been spinning. Like an Ice Skater making himself smaller to make him spin faster, the earth spin would have increased when the massive hole was made.

This increased spin would have made the effect of gravity less and everything would begin to get larger.

Biblical Evidence

In Genesis, it indicates that god placed a firmament or barrio between the waters or planets. One interpretation of this passage is that when earth and Mars came together for the last time, huge chunks of both planets were pulled into space and became a firmament [asteroid belt] that was between two sets of planets. Therefore, we can correctly determine that all this happened as the Mesozoic Age erupted on our now smaller planet. The image below shows the section that was yanked away. This is sometimes referred to as the Line of Demarcation. When it happened, it was much larger than it is today as the earth is trying to heal itself.

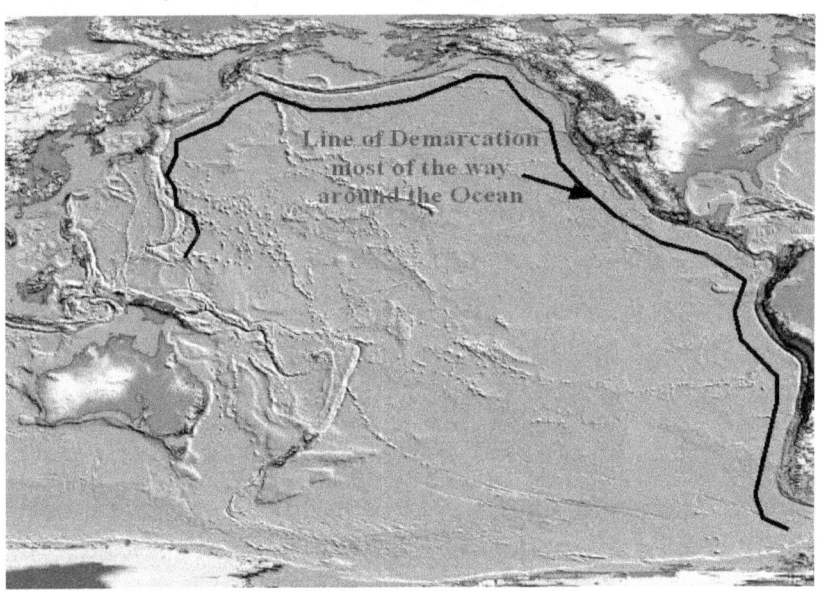

A Scientific Overview

Let's determine a "Scientific version of the Earth creation that uses details of the mathematical and physical evidence collected to date.

1. A planet called [Earth] was in a widely eccentric path, that put it close to the planet Mars on occasion. **[Not the stable orbits you had believed.]**

2. On Earth and Mars lived many creatures.

3. During at least 2 of the flybys, the Earth's major mountain ranges were produced. [**Not Plate tectonics**]

4. Once, Mars came too close and a portion of the land, where the Pacific Ocean is, was pulled away from the Earth.

5. Many creatures were killed on both planets. We can believe there were no survivors on the Martian surface.

6. The portion of the earth that was pulled away shattered and the pieces obtained an orbit around the sun to become the Asteroid belt.

7. Many pieces hit Mars and caused great craters on only one side of the planet. The other half of the planet had been ripped away during the encounter. It is believed that some level of atmosphere still remained, but it was not to last.

8. The major piece of the earth was cut loose from its former orbit and took a new orbit around the sun and Mars also seemed to have its orbit made more regular from the encounter.

9. One piece from the explosion stayed with the earth and became the moon.

10. Violent storms and floods initially filled the site of the rupture and it became the Pacific Ocean.

11. Pangea split apart to fill the rupture and the filling is still occurring today, as the Atlantic Ocean gets wider.

.

Dating the Mars Encounter

As I previously mentioned, you were probably told the Pacific Ocean incident occurred only about 212 million years ago. Certainly, with a 12 in it the number must be right. To test that answer we need to look at dinosaurs. The period from the Jurassic Age until the end of the Cretaceous Age is known for one thing—dinosaurs—BIG ones. At the end of this time, almost overnight, things got much smaller. There is a reasonable probability that one of the initiators for the large dinosaurs was the creation of the Pacific Ocean and the end of the age was quickly climaxed by the slowdown of the earth's rotation. I know those claims are not presented in other works, so let me start by listing things we think we know.

Earth Changed At the beginning of the Triassic

If nothing else, the Permian-Triassic boundary is distinctive. Something happened to put a hole where the Pacific Ocean is today and this same event caused the super-continent of Pangea to begin to pull itself apart, and killed most of the living organisms on the planet.

- According to data collected from the Deep-Sea Drilling Project [1968 to 1983] the **Pacific Ocean Basin** was determined to be youthful. This was determined by testing mantle depth and sedimentation over the ocean floor. The mantle is slowly repairing itself, but there is still a huge difference between the mantle thickness in the Pacific Ocean and that found on the rest of the world.

- According to many models, the **break-up of Pangea** occurred around 200 million years ago. One simple test was initiated to test this date. The Atlantic Ocean is the major separating line of the super continent and it is getting wider by about 2 to 3 centimeters per year with a current average width of about 6,000 kilometers. Therefore, if everything remained constant throughout this whole period, the close encounter and the explosion that made the Pacific Ocean would have occurred around 200 million years ago. The problem is that while the Atlantic is getting larger every year by 3 centimeters, we can believe that there are less forces pulling Pangea apart today, so the time period could be MUCH, much shorter.
- Major **meteors** hit at the end of the Permian period. So far, we have categorized at least 5 major craters that occurred on the Earth during that time. The craters can be found in Beaverhead Montana, Acraman, Australia, Chad, Africa, and Siljan, Sweden. The craters are so separated; it is like the initiator of the meteors was very close to the earth.
- Talk about destruction! That was probably the worst destruction ever. Over 90 percent of all creatures became extinct.

The date of the Pacific bottom, the date of the Pangea split, and the date of the most extensive extinctions point to the most traumatic event of our earth's history. That event was the encounter with Mars.

Background of Mars

Before we get into discussions about living on Mars, we must first look at things that might be considered different than here.

- Only half the circumference of earth, Mars currently has a day that is almost identical to the current earth day. Given the much lower gravitational pull of the planet, this may be too fast to hold on to an atmosphere and water like that found on Earth so why is there so many indications of water? Hint! It didn't happen millions of years ago.

- One thing strange is that its mass is only 10 percent of our planet so the massive molten metal dynamo that builds out weather patterns and shifts our planetary axis is not disruptive on Mars.

- Additionally, people would weight only about 1/3 what they weigh on earth today.

- Mars is currently much less massive than any other planet, except Mercury and Pluto which have been generally considered to have been escaped moons. Why it is located where it is suggests it once was substantially larger.

- The Martian southern hemisphere is peppered with craters. Few are seen in the north.

- The above crustal dichotomy is almost a perfect circle. This suggests that an extremely close cataclysm.

- The smooth crust of the northern hemisphere is only about a kilometer thick, compared to 20 kilometers in the south. This indicates that the northern half of the planet has been ripped away. While this fact is unmistakable, some choose to ignore it.

- The nearby asteroid belt is clearly the remains of a planetary mass that once was spinning around the sun. The problem is there isn't nearly enough mass to have been an entire planet. [We now know what really happened.]

- The huge erosion patterns on the surface indicate massive flooding in the past- [the very recent past, less than 50 thousand years ago.]

- Many geometrical shapes with multiple parallel and perpendicular lines can be found. If these were found on earth, people would indicate that they were the remains of an ancient city or cities.

- Long straight lines that go one for miles just like roadways.

- Areas that appear to be burned. The areas look like burned out cities in the aftermath of a violent war.

- Reduced Iodine-129 [Xenon-129] can be found in abundance. This can indicate only one thing. Some types of "nuclear events or massive explosions" were close to the Martian surface. [That sounds like war. Here's another question why would a war be fought unless people were on the planet?]

- The Mongulala History indicated that the war that destroyed Venus destroyed the civilization on Mars as well. [The Mongulala are in Brazil and there are lots of trees there. I know that doesn't mean anything as I'm just messing with you about the trees.]

In addition to the anomalous physical features and thanks to many satellite pictures, we can be fairly certain that life was on Mars many thousands of years ago. We will look at a number of them later. We also can be fairly certain that people of earth greatly affected the Martian development. This thought does not violate scientific reason. It simply means there was a very ancient group of humans that we know very little about. This ancient group of humans probably went everywhere including Mars. Because people are the way they are. There soon were many wars including wars between planetary colonies. Much of the destruction found on Mars did not come from it splitting apart. It also did not come from the planet slow loss of atmosphere and water.

I know I'm shifting around a bit here, but I think we need background before we get into the meat. For instance, let me give you a quick overview about how ancient humans designed and flew flying machines called merkaba by the Babylonians, vilaxis by the people of Atlantis, wheel inside a wheel by the Jews, and, most importantly, Vimanas and Vantras by the people of India.

The end of the majority of the dinosaurs appears to have been only about 120 thousand years ago and the events that made the Pacific Ocean, moon and reduced the size of mars possibly happened only about 300 to 400 thousand years ago.

Mars Mess

The 4th planet from the sun which should be named Mars-plus as it is much larger than today's planet, comes near the earth that is larger as well. As it passes, the Erath crust is pulled upward into the sky creating the Himalayan Mountain Ridge. We can believe there was massive destruction as the earth environment was rocked by the intrusion. This event triggered the Permian extinction and millions of animals were destroyed. I like to think the animals on Prestonia were more evolved than those on Pangea, but we may never know. As Mars, Venus, and Earth all fall within the delicate region that could sustain life, we can assume some type of life was lost during the horror we have just looked at.

University Modeling

New math models were able to capture the events that caused something quite different that mountains being pushed up. Instead, the mountain ranges had to have been **pulled up**. A large planetary object strafing the planet made each of the extended mountain ranges. Once the Earth was strafed with the Earth rotating on an axis through the middle of the Pacific Ocean and Asia and a second time when the rotational axis was similar to our present rotational feature. Yes, I did say the earth's axis changed. In fact, we will see that it has happened more than once as was shown in the section on Archeo-magnetic timing.

Mesozoic Life

The next thing we need to examine is when people became civilized. This section tells of another "vain truth" taught that presents the first "modern man" as Cro-Magnon who entered the scene about 40 thousand years ago. They know there is strong evidence of the previous people, but it doesn't "FIT" in their version of the truth. Rock strata from Triassic, Jurassic, and Cretaceous periods contain literally hundreds of thousands of dinosaur tracks, and actually outnumber bones by orders of magnitude. After all, dinosaurs only made one skeleton and many footprints in its lifetime, so we can get a better understanding of dating from the footprints.

Tracks Found Everywhere-Dinosaur tracks have been found in over 1000 locations throughout the world, on every continent except Antarctica. In the U.S., they are especially abundant in Texas, Colorado, Utah, Arizona, New Mexico, Connecticut, Massachusetts, and New Jersey. It is believed that in the western U.S. alone new sites are being reported at the rate of about 50 per year. Most of these tracks have been found where there once were shorelines large expanses of moist sediment were so important in building proper fossilized tracks. Here is the weird part. Human footprints are being found with the dinosaur ones. Many times, these footprints show humans that were huge lived with the huge dinosaurs.

- *At **Rocky Hill, Connecticut** can be found a track floor that is covered with hundreds of theropod tracks.*

- *In **Amherst, Massachusetts** one can find thousands of lower Jurassic dinosaur tracks from the Connecticut Valley of New England.*

- *At **Glen Rose**, Texas one can see many large Cretaceous carnosaur and sauropod tracks still in their original positions.*

- *At **Seneca, New Mexico** the site contains hundreds of ornithopod and theropod tracks.*

- ***Tuba City Arizona** has a site containing many lower Jurassic theropod tracks.*

- *In **Denver Colorado** several Cretaceous dinosaur trackways can be seen still in their original position.*

- *At **Alberta, Canada** a vast collection of dinosaur tracks from the Peace River of British Columbia can be seen along with one of the largest exhibits of dinosaur skeletons.*

- *In **Price, Utah** one can see displays that include about 50 Cretaceous dinosaur tracks collected from coal mine roofs.*

On and on we could go, but the thing that is unusual is that many sites have human footprints mixed in. The graphic following shows some of the trackways that have uncovered the evidence of people walking on the same beaches as dinosaurs.

Turkey-In the late 1950's during road construction in Homs southeast Turkey, many tombs of Giants were unearthed. These tombs were 4 meters long, and when entered in 2 cases the human thigh bones were measured to be over 47 inches in length. It was calculated that the person [or Titan] who owned this Femur probably stood at **fourteen to sixteen feet tall.**

Mexico 1925-According to the Washington Post, June 22, 1925, and the New York Herald-Tribune, June 21, 1925, a mining party found skeletons measuring 10 to 12 feet, with feet 18 to 20 inches long, near Sisoguiche, Mexico. These also sound like Titans.

In South Africa, a giant footprint of a woman measuring over 4 feet long has been dated to be from before or during the Triassic period. Pointing to the probability of this being a female human-like species' foot, proportionally the two-legged being would need to be <u>some 30 feet tall!</u>

Chinese Titan Footprints-In Shenmu County in China's Shaanxi Province, in 1967, a man surnamed Qiao went to quarry some stones and found huge human-like footprints encased in the stone heading to the edge of a cliff. Each of the footprints is about 16" long indicating the human was about 9 feet high. Considering the footprints are embedded in the stone, they were determined to be Cretaceous.

Giant Bones in Ecuador- A find from the estate of a Catholic priest in Ecuador points to the existence of Titans in Ecuador. A collection of giant fossilized bones was found in the estate of the late Father Carlos Vaca. Anatomists identified one of them as the occipital section of a human skull. They believe that another bone may be part of a massive human heel. Judging by its size, it would have belonged to a giant human that may have been over 7 meters tall [over 20 feet tall]. The bones have been dated to be Cretaceous.

At Inverell, Australia-we find an example of the Titan evidence long, long ago. The footprint below right shows the tremendous size that some of these people got. This particular one was found along with many others.

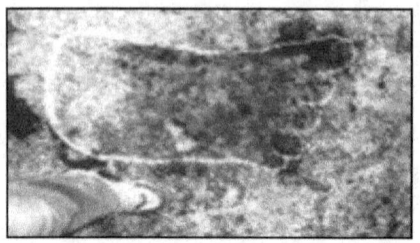 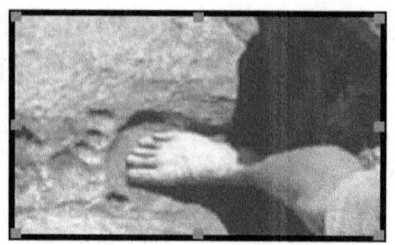

Australian-In September 1993, another giant-sized human fossilized footprint was added [above left]. Here, embedded in the rock, was a large footprint impression. The track was that of a right foot, probably distorted in the original soft mud, and was 44 cm in length x 30 cm across the toes. There were signs that other tracks had been embedded nearby, but these had weathered away. The imprint was preserved by a lava flow that was "reportedly" dated about the time of the Homo-Erectus. The monstrous human whose single footprint still survives was about <u>11 feet in height</u>.

Giant "Human" Shoes in Australia- Don't believe these people only walked on the beach barefooted. The next photos are from NSW, Australia of shoeprints. Next to the shoeprints are images of the photographer's "tiny" shoes. As shown to the right is the general placement of the 2 complete shoe prints and a number of partials that were found. Now, when I found these prints, there were a number of prints and half-prints, which appears to be three sets of different prints, interlocking with each other. Don't ask me why people wore shoes on the beach, but there you have it.

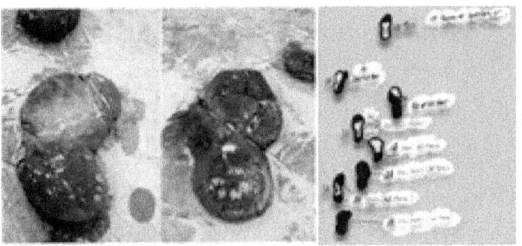
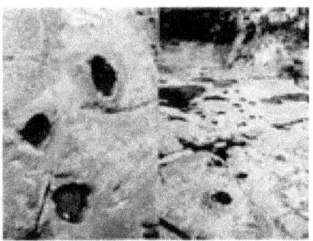

Another group near this same location also appears to be human shoeprint, but this time, the raised edges show that he was getting stuck in the mud as he walked possibly from the Cretaceous time as well.

Skeletons

Arizona-In 1923, Mr. Samuel Hubbard discovered the remains of giants in the Grand Canyon of Arizona. The discovery consisted of the following: Petrified bodies of two human beings about <u>18 and 15 feet in height</u> respectively. One of these was buried under a recent rock fall which required several days' work to remove. The other, of which Mr. Hubbard took photographs, was in a crevice and inaccessible. The bodies were formed from limestone petrifaction and embedded in sandstone during a very ancient time.

Nevada-In July, 1877, four prospectors were looking for gold and silver outcroppings in a desolate, hilly area near the head of Spring Valley, not far from Eureka, Nevada. One of the men spotted something peculiar projecting from a high ledge. The prospector was surprised to find a human leg-bone and knee cap sticking out of solid rock. He and his companions dislodged the oddity with picks. Realizing they had a most unusual find, the men brought it into Eureka, where it was placed on display. The stone in which the bones were embedded was a hard, dark red quartzite, and the bones themselves were almost black with carbonization showing its great age. When the surrounding stone was carefully chipped away, the specimen was found to be composed of a leg bone broken off four inches above the knee, the knee cap and joint, the lower leg bones, and the complete bones of the foot. Several medical doctors examined the remains, and indicated that they had indeed once belonged to a human being, and a very modern-looking one. But for us the best part was their size: From knee to heel, they measured 39 inches. Their owner in life had thus stood <u>over 12 feet tall.</u> Compounding the mystery further was the fact that the rock in which the bones were found **dated to the era of the dinosaurs**, the Cretaceous or earlier. The local papers ran several stories on the marvelous find, and two museums sent investigators to see if any more of the skeleton could be located. Unfortunately, nothing else but the leg and foot existed in the rock. Again, this is just a tiny sampling of huge amounts of information.

Ancient tests tell us most about what we know of this ancient group. We understand that they had great cities and a high level of civilization. They also had scientists and modified animals making some of the largest and most odd looking "mistakes" ever seen. About 120 thousand years ago, according to the book of Isaiah, Jerimiah, and Genesis, the entire world became desolate and all the cities were completely flattened. Many ancient texts call this time, "*The War between the Giants and the Gods*". Titanic people had been the undisputed rulers of the land for hundreds of thousands of years and all was gone. While almost all the physical evidence of the civilization of Titans was lost with time, occasionally things do pop up. Here are a few things that show mankind advanced to a point that would allow travel to the Moon, Venus, and Mars.

Mesozoic Civilization

These ancient people were not lumbering fools as some of the Greek histories try to portray. They were very civilized and skilled in all types of science and technology. As such they produced thousands of tons of those nasty hydrocarbons that are being talked about today. Here is a sampling of some of what has been dug up showing their advanced levels on manufacture and science.

Building Materials and Art

Iowa-1897-A large stone [2x2x1feet] with multiple faces of an old man carved on it and a grid pattern on the remaining area was found 130 feet down in a coalmine. The estimated age was Cretaceous. [Below left]

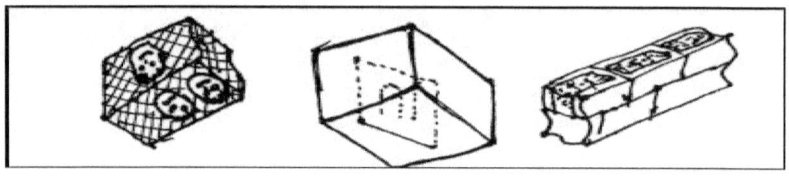

Philadelphia 1829-A 30 cubic foot piece of marble was excavated from a depth of 60 feet. Inside the marble was a straight edged rectangular indentation. After a section of the marble was carefully removed, it was found that 2 distinct heavily engraved letters similar to an "I" and a "U" eleven inches long and 5.8 inches deep were on a square base. The estimated age was Cretaceous.

Oklahoma-1928- [Above Right] A block wall was found almost 2 miles deep in a coal mine. Each block was 12 x 12 x 12 inches polished on the outside and filled with gravel on the inside- There were multiple reports over 150-yard length of the same wall. The estimated age of the wall was Cretaceous.

Strange Geode

The old battery in a geode-The picture next is some type of power conversion device found **inside** a geode, in California. Below the geode is a graphic of x-rays of the geode showing the elemental parts. These include a spring, core, plate, and electrical insulator. The same parts as you would expect in a battery. Maybe this is a new way to package batteries, but it takes a long time to complete the package. Both of the objects are extremely ancient and certainly, before we originally thought that everyone used electricity. The central metal core surrounded by the white material looks like a battery. Whatever it was, it was electrical. On the right is a drawing of the parts and a size comparison to a standard D-cell battery.

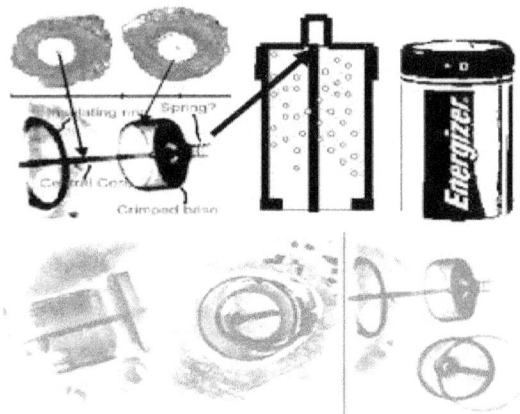

Ancient Nuclear Plant- Surprisingly there are many more ultra-ancient artifacts that show a high level of civilization by this early group, but I think you get the idea. The last thing I want to bring up is the massive nuclear plant in Oklo Africa. A nuclear plant with at least 16 different areas of uranium depletion has been found and this massive thing is dated to before the end of the Jurassic Period. The plant complex is unusual in that there is no nuclear fallout. I know you were thinking these people were so backward, they had to rely on simple batteries for power, but I'm talking about nuclear plants, bombs, nasty wars etc. While built during this time, these plants were, evidently used over the years by others. In Oklo, Gabon, Africa, we found 16 depleted uranium pockets or processing areas inside some caves. Very quickly, scientists backpedaled and came up with the story that this was a naturally occurring nuclear plant, just like any other natural nuclear plant. Wait a minute!!! There aren't any! These nuclear reactors were estimated to have produced on the order of 1,000 megawatts, comparable to a large modern plant. All this could have been used to light houses for war or even to fuel space vehicles.

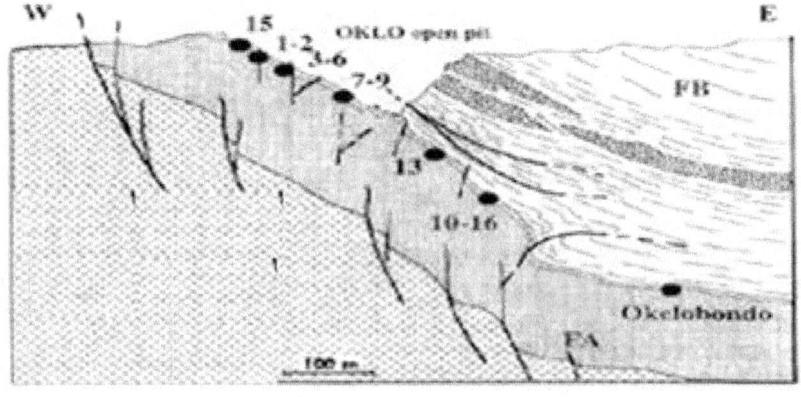

The normal Uranium 238 was "processed" to Plutonium and "enriched" to Uranium 235 allowing energy to be provided and used [by someone]. The following graphic shows where the processing took place. I know it just looks like blobs, but believe, me this was put to use in the olden days and we have proof to look at later as it would not just be used by the Titan people, but also the group called the Anak people along with the Cro-Magnon people of the Pleistocene Age that would initiate a war that would eventually lead to Venus being destroyed in an even worse way than the Mars horror.

It had been about 300 thousand years between when Mars and the Earth had both split and a massive war erupted. We know quite a bit about this war because most of the dinosaurs died, at the end of the war some massive object hit the earth causing a temporary shift in the earth axis and on the other side from the massive blast, the earth was split open and spewed out magma for many years. Let me go over some of the data we have, but in the back of your mind be thinking about how a 300-thousand-year-old civilization may have been able to populate the once decimated ½ of a planet we call Mars and the much closer and lush planet called Rahab in the Bible. Right now, we have to fight a war and end the Cretaceous Age to end the Mesozoic Period and destroy most of the Dinosaurs.

Cretaceous War

Most know about this iridium chalk layer that is found at the stratographic position determined to be the end of the Cretaceous. According to many, this comet from outer space hit the Yucatan and blasted a hole in the Earth. It hit so hard that the comet exploded and send iridium all the way around the earth marking the end of the dinosaurs. Some of that seems reasonable. Whatever happened, the Iridium in the boundary layer can also been attributed to another source besides a meteor. The earth's core could have done it if the core somehow got to the surface. That is where India comes in. The entire country of India wasn't here before the K-T layer was formed. Whatever hit the earth near the Yucatan caused the earth to split open and start spewing out magma. This continued until a newly developed landmass was formed that we call India. The air was filled with soot. Today, hundreds of cubic miles of magma <u>still fill the area called the Deccan Traps</u>. [Yes, I said MILES] By the way, dinosaurs weren't all killed by the Chalk as some have indicated. Many, many dinosaur fossils have been found <u>below </u>the K-T layer showing they were killed a thousand years or so before the Chalk event, DURING some massive nuclear war. <u>Many Mesozoic Age dinosaur bones are radioactive,</u> but most areas where they are found are not. During the Mesozoic Period of the dinosaurs, scientist readily admit that the Oklo nuclear plant was running, but no one wants to say is if animals were radiated, nuclear products must have been in use. At the same time, the Bible indicates the war that killed most of the dinosaurs, and left the entire world without form must have been a hum-dinger as it made many of the dead animals radioactive.

Isaiah 14:16-17-Is this the man that made the Earth to tremble, that did shake kingdoms. That made the world as a wilderness, and destroyed the cities.

Jeremiah 4:23-27- [near the end of the wars] I beheld the Earth, and, lo, it was without form, and void; and the heavens, and they had no light. I beheld, and, lo, there was no man left, and all the cities thereof were broken down

*Nag Hammadi-The heaven and Earth were destroyed by the troublemaker that was below them all. The sixth heaven shook violently- when Pistis [God] knew about the breakage, **she** sent forth her breath and bound him [Satan] and cast him down into Tartaros.*

One thing that is known by scientists; while Iridium is only found on meteors and deep inside the earth, the amount of iridium found around the world in the K-T layer is far too dense to have come from a single meteor. The reason they don't tell you is that most of the iridium came from the Earth and the country of India tells the awful story. The graphic following shows how chalk killed the giants, sent massive amounts of hydrocarbon blasts into the Ozone, and eventually created the country of India.

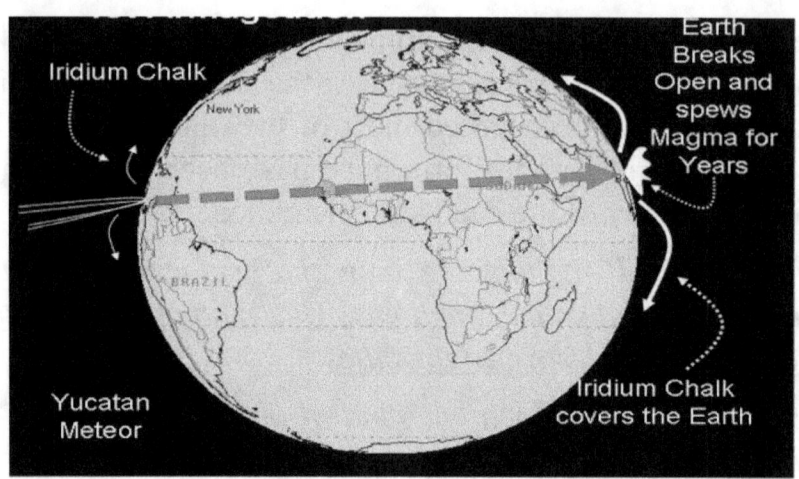

When the massive meteor or "whatever" hit, the other side of the world, belched forth millions and millions of tons of magma from a massive opening that formed. This huge pile of magma is called the Deccan Flats. Originally, this massive pile of "earth insides" covered over 2,000,000 square meters and contained about a million cubic kilometers of magma rich in iridium. This sort of thing has happened before man was on the Earth, but this time was the only one since mankind was created and it seems to have caused the destruction of the Titan people if any remained after the Great War period that ended what the Greeks and PreMaya called the Golden Age. The Titan People were all gone, but the Earth remained solid.

Leaving the Planet

While there was horrible fighting on the earth, we can believe some left to start new colonies on Venus and Mars. When we get to the section on artifacts, remember this opportunity. Artwork shows they knew about flight, they had electricity and nuclear power--- and, most likely they would not like staying here are the earth was almost completely destroyed. Let's say some people did not want to be on the Earth during the time of the nasty war that ended the Cretaceous. When the Mars colonists in ancient times left on their trip, they probably never returned. I would assume they originally thought they would be away from a war torn and greedy planet. When people got to Mars, they built structures and rural areas and cities. Unfortunately, we know the wars would not leave them alone.

Evidence strongly suggests that the nearby planets of Venus and Mars both were colonized. We are going to look at a smattering of the evidence that has been found by many researchers in this area. There is little doubt that people once lived on both planets and both have substantial signs of war. The hard job is to determine when the colonization occurred, when and if colonization ceased, and why civilizations ended. On Venus we get lucky when determining the end of civilizations. That ending occurred 11 thousand years ago. We know this for many reasons, but to find out what happened on Mars will be more difficult because the catastrophe was much less drastic.

This first major war was called by many names. The Jews called it the 1st Heaven War. This guy named Satan essentially destroyed both Heaven and Earth in the process. Here are some of the descriptions. Chapter 6 of Genesis tells us that the people living during the very ancient times before Adam were known as the great men of old. The 1st chapter indicates that God had to remake man to RE-populate the world after it had become void and without form

Greek Myths

Titans battled the creator. Aristophanes wrote, *"Terrible was the might of the **first beings**. They had proud thoughts and **made an attack on the gods.**"*

New Zealand

According to the Maori tradition, *the sons of Rangi and Papa were not unanimous in the decision to separate their parents [split the heaven and Earth apart] so a huge war of the gods followed the separation. After a 2nd war in heaven, Tane forced rebels <u>to other worlds</u> of darkness and despair.*

Egyptian Text

"Book of the Dead"*-**The Gods Osiris, Seth, and Horus had a long-lasting war. According to the ***"Creation Epic"*-The forces were not conquered forever at the beginning of time. The **war between darkness and light** sustains the world; and when it comes to the final end, so too will the world.*

Magan Text

This comes from Ancient Babylon- *"The ancient ones that bore all the waters were one. Rebellion arose in heaven. Absu rose up to slay the Elder God, but was slain."*

Tibetan Text

***Book of Dzyan*-**Tibetan religious documents indicate that *God's sons were told to make an image, but **1/3 refused and battles were fought** between the creators and the destroyers.*

North American Tradition

***Paynut Indian tradition*- ***"God Hinuno **battled the other gods** and some gods were thrown out of heaven."*

Civilization

Some suggest that the war could not have been devastating, that technology was not to a high level, but we now know that electric power, at least in the form of batteries were available, all types of construction tools were available, people wore similar clothing to what we wear today, and the people living during this time looked like us. Many were much larger than we are today, but, certainly the physical characteristics were the same. We know these things because we have found physical evidence all over the place.

How did We Find the Mesozoic Nuclear Plants?

In 1972, French scientists discovered that several "supposedly natural" concentrations of uranium ore had become critical and flared many thousands of years ago at Oklo, Gabon. They discovered six ancient zones of depleted uranium with plutonium byproducts and that's no easy task. The mine's nuclear reactor was several miles long and said to be well designed, whatever that means. The radioactive wastes from the reactor are still confined inside the mine itself and analysis of the waste demonstrated that **plutonium had also been created**. Some scientists believe that the mine was in existence before the Mesozoic, but it is very difficult to date nuclear events because of the change in atomic structures that occur in the entire surrounding. I tend to believe that this power source was the remains of that used by the ancient humans. If nuclear power was used, then more than likely, it wasn't just used for good things during this very ancient war time.

When Was The War?

My assumption of the first war timing is that it occurred at the end of the Cretaceous, 120 thousand years ago and there are many reasons to use this timing.

The destruction and change of the earth as the Cretaceous Extinction destroyed most living animals is completely different than those before or since.

The earth evidently could no longer support the animals from before this time.

The signs of civilization became less after this critical time period.

Over 85% of all life was destroyed at this time.

Reptile dominance ceased after this destruction time.

A new primitive man came into existence after this time as marked by the Genesis story.

No destruction or extinction to that level has occurred since that time.

The slowdown of the Earth insured that not more of the huge animals could be made [I know you probably have not heard about this important fact but there are a number of things our historians seem to leave out of our text books.

Slowing Down

As I mentioned, Mars yanked out a part of our planet about 400 thousand years ago and this caused the earth to spin faster----gravity became lower---- our atmospheric gases began to get less and less as the atmosphere could not hold on to them-----and just about everything on the planted got bigger to compensate for the lower gravity. Gravity on the Earth was lower for a time and the great beasts that the

Sumerians indicated were bred for the Heaven War, got bigger and bigger.

Split Open

One of the events at the end of the Cretaceous that brought the Heaven War to a close and helps us date the war is that a massive boulder from space hit the Yucatan. As I stated, it hit so hard that the other side of the planet ripped apart. Million and millions of tons of magma spewed for months or years and the country of India was made. [I know you were told it was a little sliver of a plate that rammed into the Asian continent, but the Deccan Flats tells us a completely different story. The picture below shows one of the largest piles of lava. It is in India, in fact it used to cover just about the whole country and it still covers over 200,000 square miles and is 6,500 feet thick. It is estimated that originally it covered 2 million square miles and it still contains over 12,000 cubic MILES of lava.

*[**That's right I said MILES, not meters**] When the Earth splits open it splits wide open.*

While the today this huge magma bed only covers the Dark section shown below, when the split happened, almost the entire country was made up of Earth guts as shown as the larger area.

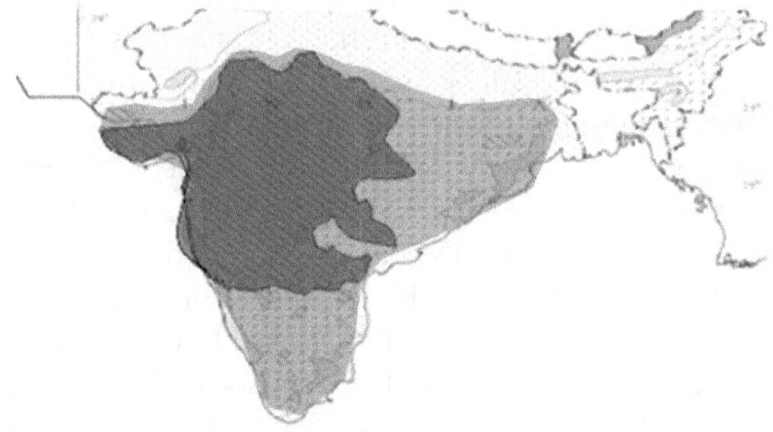

The war and, of course the big boulder hit, essentially destroyed the Earth. Many died. Cities were completely destroyed. The Bible gave it bad marks.

Formless & Void

Darkness was on the face of the deep. *[Genesis 1:2]*

This talks about the desolation of the whole Earth from huge battles. The Biblical book of Jeremiah provides insight about what is meant by without form and void. Following are some excerpts from different Jewish histories that all say the same thing. Some huge war in the physical world left things unpleasant.

Genesis 1:2-And the Earth was without form, and void; and darkness was upon the face of the deep. **[Note how the Book of Abraham completes this verse.]**

Book of Abraham 3:27-And the second angel [We know him as Satan] was angry and left heaven and many followed him. Then the Lord came down. They went down at the beginning, and they organized and formed the heavens and Earth. The Earth, <u>after it was formed</u>, became desolate, because they had not formed anything but Earth, and the spirits of the gods were brooding upon the face of the deep. **[Clearly this "turning earth to a formless void" was after Satan, the angry second angel, and those that followed him left heaven. The brooding gods were the losers of the war.]**

Isaiah 14:16-17-Is this the man [talking about Satan again] that made the Earth to tremble, that did shake kingdoms. That made the world as a wilderness, and destroyed the cities. **[Notice that it indicates there were cities before the world was turned into a wilderness.]**

Jeremiah 4:23-27- *[near the end of the wars] I beheld the Earth, and, lo, it was without form, and void; and the heavens, and they had no light. I beheld, and, lo, there was no man, and all the cities thereof were broken down. For thus hath the LORD said, the whole land shall be desolate; yet will I not make a full end.* **[This is discussing this first war, it shows that a high level of civilization was thriving before the war.]**

Mars And The War

While we don't know much about how any colonists of Mars or Venus might have aided in the war, the second chapter of Genesis and other verses appear to tell us that the people on these Planets were involved in this war and the others we will briefly examine. The Bible called the planets the "waters" or "life giving places". I'm not trying to interpret into a definition here. There is no way the "waters" interpretation can mean actual water in these verses and if the planets were involved, there must have been people on the planets. The ones most reasonable would be Mars and Venus. Let me show you what I mean.

Genesis

Genesis 1:2 "Then the Earth <u>became</u> without form and void. Then <u>Darkness</u> was on the <u>face</u> of the <u>deep</u>. And the spirit of <u>Elohiym</u> moved on the <u>face</u> of the <u>waters</u>." [That word "DEEP" by the way did not mean depth of the ocean as you would imagine. There was a different Hebrew word for that and there and Hebrew isn't like English where a hundred words might mean the same thing. This deep was the emptiness of the heavens and the word waters was not water like drinking stuff or oceans stuff. There were different words for those. Waters meant "life giving place" or inhabitable Planet. This was talking about after the WAR and the losers "called Elohiym" here, had to find other places to live.]

"Jeremiah"

"Jeremiah10:13"- *"He hath made the Earth by his power, he hath established the world by his wisdom, and hath stretched out the heavens by his discretion. When he uttereth his voice, there is a multitude of waters [**inhabited planets**] in the heavens."* [Please don't believe that little globs of water are what are being talked about here. Different than illuminations [Stars], these "Waters" could sustain life.]

"Psalms"

Psalm 148: 2-4- *"Praise him, ye heavens of heavens, and ye waters/ [**inhabited planets**] that be above the heavens."* [The only reason a "waters" should praise God is that they are talking about the people living on the planets, not a liquid.]

"Book of Abraham"

This comes from the ancient Jewish book of the Essenes ***"Abraham 4:6"-****"And the **gods** also said let there be an expanse in the midst and it shall divide waters from the waters, and the gods ordered the firmament so that it divided the waters which were under the expanse, from the waters which were above the expanse.* [The waters above the expanse were planets. The two most colonized were Venus and Mars.]

"Origin of the World"

This comes from the Jewish/Egyptian Gnostic work "Origin of the World". *"Then the bile [Old Satan] that had come into being out of the shadow was thrown into a part of chaos. Since that day, a watery substance has been apparent. It appeared as a spirit moving upon the waters.*

And when that spirit appeared, the ruler set apart the watery substance. And what was dry was divided into another place. And from matter, he made for himself an abode, and he called it 'heaven'. And from matter, the ruler made a footstool, and he called it 'Earth'. [Seems to be saying that the planets were populated AFTER Satan lost his war.]

"Apocryphon of John"

Another Gnostic Test "Apocryphon of John" provides more information. *"And the "arrogant archon" [**SATAN**] took a power and when he saw the multitude of the angels which he had created, then he exalted himself above them. And the whole group of archeons [followers of Satan] trembled, and the foundations of the abyss shook. And of the waters which are above matter was illuminated.* [Seems to be saying that the colonists on the planets were illuminated after the war. Possibly that people had to move to the nearby planets because the Earth was in such rotten shape.]

Mars and Venus Inhabited

As these verses tell us, during a very ancient time, the inhabitable planets were colonized. One of these planets was Mars.

Tertiary Age

The Tertiary Age by the new timing lasted from 120 thousand years ago until about 40 thousand years ago. As the Titans disappeared, the ANAK People came along to control the world. According to the Bible, these were actually the Titan People who had died. They fought a horrible war and lost only to return as people to a devastated earth. We can imagine that this could be referring to those colonizing Mars and Venus as the war waged on. I don't want to get into this in great detail as I need to get to details of Mars. Certainly, there are many other sub-races, but the Anak were the dominate people of the Tertiary Age. These people were almost as large as the Titans and they lived during the Tertiary and past Pleistocene Age. Many died at the end of the Pleistocene as another great extinction showed up in the Ice core sampling, but before we get to that let's look at radioactivity.

Radioactive Dinosaurs

We know the ancient people had the resources of the Oklo Nuclear plants and they, evidently used nuclear material in some bad way as unfossilized dinosaurs are now being found that are so radioactive, they must be painted with a heavy lead base to reduce the emissions for display in museums. Some try to say as bones are mineralized, they suck up radioactive materials and become more radioactive than the ground so people go around finding dinosaurs with a Geiger counter.

Leaving the Planet

There are specific references to colonization of the planet Rahab [Venus] in the Bible and other historical references, but we can assume that there was still enough air for survival even at this late date. It is believed that the Martian colonists would have been forced underground as the air had greatly thinned by this time. Possibly it wasn't so bad as they left Earth in terror to find new hope in Mars. To be sure unground living would change people. We can imagine they would gain larger eyes to support subsurface lighting and we can believe most of the oxygen at this time was "manufactured" therefore it would have been much more oxygenated that Earth atmosphere so the nostrils would have reduced in size as would the lungs.

Look at the Horrible Wars

We are told of three major wars established on Earth. I'm not talking about the tiny World War II we had which caused the death of less than 1% of the population. I'm talking about nasty wars that affected the ENTIRE world. The wars were so bad and widespread that we can assume colonists living on Mars would have been involved. According to ancient Judeo-Christian texts and other writings, each of these major war eras resulted in the population of plants, animals and humans to be reduced by at least 1/3. To keep from being one of the "killed" ones, many began living underground. Around the world every year we find more underground cities and settlements from the time. From written descriptions and artifacts of the wars we can determine the timing of the wars.

Second War

To find out when the second war happened, we go back the ancient texts. Biblical books of Jubilees and Jasher and Job paint a pretty good picture, but other texts also help us understand what happened. Let's see what was said.

Persian Tradition

Form ancient Persian stories we find this interesting section. *"Ahriman [Satan] tried to destroy the world again and turned the Earth into a desert, killing plants and animals. Preserved seed of the plants and animals came from the moon and new life came."* This tells us that there was an output on the moon. We don't know how they got there yet, but you just started the book and will have to be patient. We also know that this is a second war that demolished everything on the earth just about. We need more information. For that we go to the book of "Jubilees".

"Jubilees"

Jubilees 5:1-9-*"And against their [Anak people] sons went forth a command that they should be smitten with the sword----And he sent his sword into their midst that each should slay his neighbor, and they all began to slay each other till they fell by the sword and were destroyed from the Earth. And their fathers were witnesses of their destruction, and after this they were bound in the depths of the Earth."* More evidence of a massive war before Noah's flood, but what did the Inca say?

Incan History

"During the age of the giants, a huge war broke out. The war between giants and gods ended in complete destruction." According to the Inca, the age of giants is the second age and, therefore, it would most likely be talking about this same horrible War. The world AGAIN was in complete destruction, but when was the war? The Biblical book of Jasher may tell us.

"Jasher"

Jasher 2:5-6- *"-and the sons of men forsook the Lord all the days of Enosh [Adam's grandson] and his children; and the anger of the Lord was kindled on account of their works and abominations which they did in the Earth. And the Lord caused the waters of the river Gihon to overwhelm them, and he destroyed and consumed them, and he destroyed the third part of the Earth, and notwithstanding this, the sons of men did not turn from their evil ways--"*

All we need to know is when the grandchildren of Enos lived and we have it. I'm not going into this analysis in this book, but what is believed today is that Adam was created 40 thousand years ago. Scientists call this new human Cro-Magnon man and he was substantially different that those who came before. This person had a soul. Forget I said the soul thing if you want because this book is about Mars. Anyway!! Adam and his descendent lived a long time. The book of Adam and Eve tells us that Adam lived 5500 years. If we know that the flood occurred 10 thousand years ago, we can determine that the war occurred sometime between about 20 and 10 thousand years ago. If we want to know some details about the war, maybe the texts from India can help us out.

"Ramayana"

"Atlanteans in Vailixi flying ships and Indians in Vimana flying ships" battled on Earth and Moon."

"Maharishi Bharadvaya"

In this work there are direct indications of gigantic battles in heaven. [In this case heaven was actually the nearby planets. Did other societies talk about this flying to the heavens?]

Babylonian Version

In the "Epic of Etana" we read, *"Etana looked down and saw the Earth had become like a hill and the sea a well and so they flew for an hour and Etana looked down and the Earth was like a grinding stone and the sea like a pot. After the third hour the Earth was only a speck of dust and the sea no longer seen"* The ship, of course, was going into outer space.

Chinese Version

Methodology of how to **send a detachment of men onto any planet** was described in ancient documents from Lhasa. These documents were found fairly recently and have been only partially deciphered. The remaining information is being deciphered as we speak, so we may find out more about the space war in the near future. Here is a no brainer. If people were sending soldiers to some planet, there probably were people on the planet.

Greek Version

From Greek legends talking about battles between the gods we are told the following: *"Hot vapor lapped the titans, flames unspeakable rose bright to the upper air [outer space], lightning blinded their eyes."* Apparently lightning weapons were used in outer space during this 2nd horrible war. If we get back to the Biblical book of Job, it might narrow down the search for truth even more.

"Job"

Job 26:12- *"The boastful Angel and his followers rebelled [again]. Yahweh destroyed their dwelling places. He divideth the sea with his power, and by his discretion <u>he smashed Rahab. It was reduced to stones of fire.</u>"* For this we need to understand where Rahab was. Whenever Rahab was destroyed, the war was ending. This verse also tells us that the followers of Satan were stationed at Rahab. Let me just say that Satan and his gang lived even longer than Adam and many of them survived the time when Rahab was reduced to stones of fire [meteors hitting the earth that caught EVERY THING on fire.] Anytime you hear the word RAHAB as a planet, the text is talking about Venus. We'll talk about Venus a little more later, but let me give you a little bit of information up front because this second war and the whole Venus thing is very important when discussing the people who lived on Mars. The image below is not a series of craters as each hole is exactly the same size. This is as if the areas was strafed from a very low altitude in a well-planned attack. We can even tell the direction of flight as the last blast hole on the right is on top of the preceding one so the attacker appears to have been coming from the left of the screen.

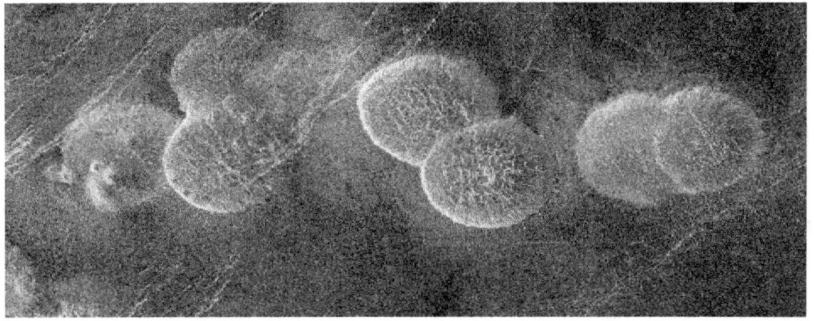

What About Rahab?

Before we go on, let's push aside Venus as the great vacation spot of the Solar System. Just about all evidence of people living on Venus was destroyed when it burst into flames 11 thousand years ago [Just before the Pleistocene Extinction] so we'll have to take what we can get. Since its initial meltdown, it has been burning ever since. In ancient days Venus was known as the wavy-haired planet, or the vain planet. Assuming the Planet named Rahab [Hebrew for "Vain One"], in the Bible was the planet Venus, we can determine that there was a horrible war at that very ancient time and the people of Venus were the eventual losers According to ancient texts, *Venus changed its position in the sky and the army of Satan was destroyed instantly as God seeked vengeance* on a second attempt to take over that Heaven place that is talked about in ancient Judeo-Christian religious texts. Venus was almost split in half from the assault, whatever that was, and it still bears the mark of a time when some horrible force just about turned our sister planet into two smaller cousin planets. This book isn't going into the Venus stories too much because they will simply confuse the issue. I want to concentrate on Mars. I simply brought this out so that no one would start wondering why the planet Mars was colonized when a perfectly good Venusian planet lay only 5 million miles from us. Just about the only ones that got away from the now burning planet were a group of people known as the Dropa.

The Dropa Family

This family or group landed in China 11 thousand years ago and recorded their trek on 716 disk-like texts. Several individuals have deciphered the writings so there is a pretty good overview left behind that tells about their entry into Chinese society. The disks told of the landing and attempts at talking with the locals. The locals killed some of them, but finally they were accepted and they became one of the "normal" groups of people. The Dropa were smaller than the locals and they had small noses and large almond shaped eyes. Some suggest that they crash landed. I contend that there was no home for them to go back to. What is reported to be remains of one of the Dropa is shown below.

We'll discuss them a little more, later, as we firm up the 11-thousand-year-old war timing a little more. If we look at Venus now what we see is the end of a planet. The next image shows a massive gash across the equatorial region where most of the massive meteoric craters are located as if whatever hit the planet and destroyed it was very close and orbiting around the planet.

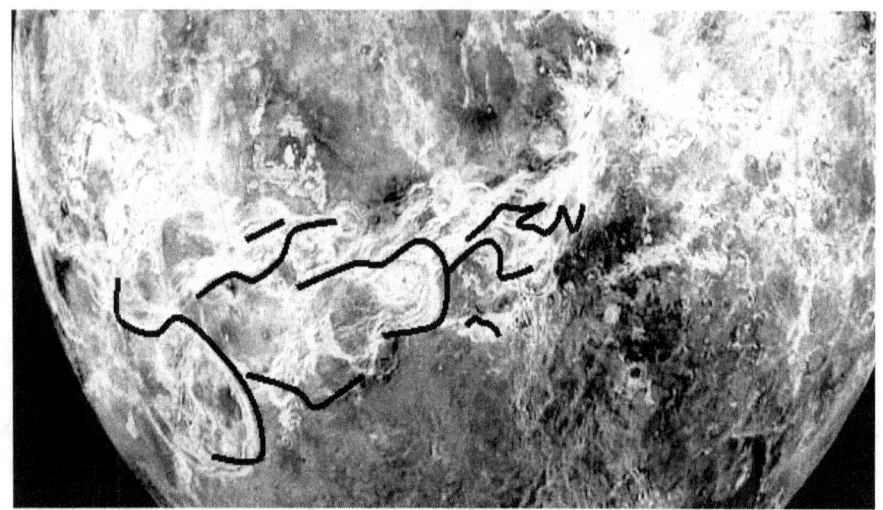

Earth would not be unharmed as some of the remains in the form of over 500 thousand masses hit the earth along what used to be the equator. The 500,000+ Carolina Bays goes from Louisiana to well past Pennsylvania. The first image below shows a small section while the second picture shows the meteors were hitting so fast that they made craters inside other craters.

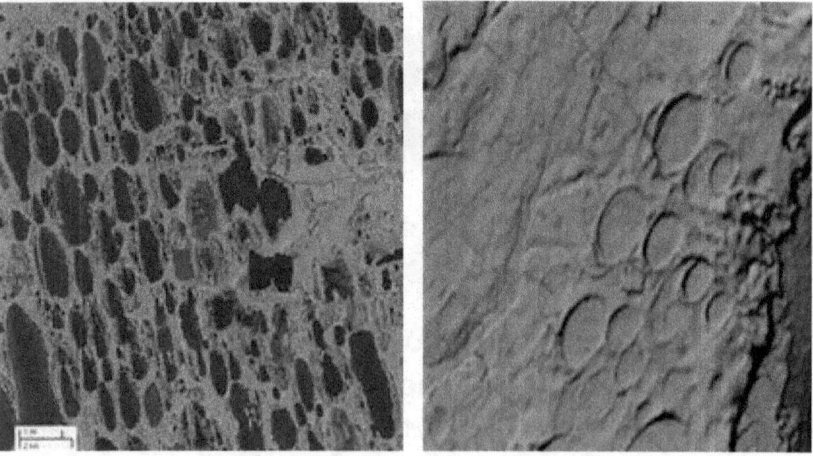

The following image shows the area where the craters are still extant.The path shows us the ancient equator and to the right we see that the Earth eventually shifter 30 degrees taking the Mammoth herds from meadow to quik free in the "New" Siberian Arctic. They would not survive it.

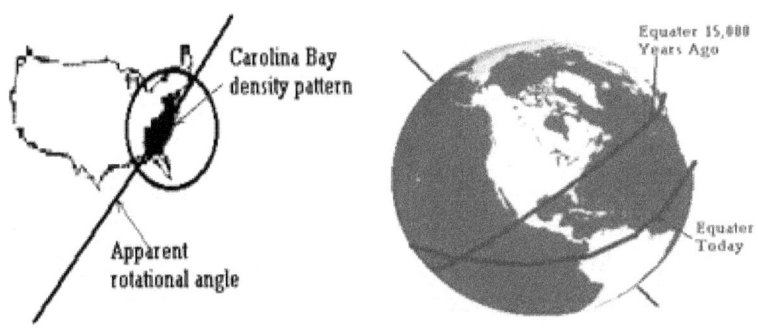

While the Venusian settlements were destroyed, we can believe Martian colonists simply looked on in horror. We can find some evidence of war on Mars but we cannot date the evidence as it is simply too far away. Possibly the limits of the war also extended out to Mars as we will see later. War or no war, the Martians had to move underground by now. As they built their society for survival, there were and possibly still are major attempts at terraforming Mars to add Oxygen and possibly even extend living outside the protected areas. For now, I want to complete the war overview with the last of the great wars. Yes! You are right. I'm talking about the last huge war that occurred 6 thousand years ago and 1/3 of the population of the entire world DIED.

War #3

Worldwide trade is easily recognized in our earth's infancy as mankind quickly spread his influence to all parts of the planet. While the very ancient civilizations could not leave much information that would last the test of time, we can see striking similarities between the ancient writing of Mohenjo-Daro in what is now Pakistan and the Easter Islands on the other side of the world. Both civilizations flourished thousands of years ago. On and on we could go as mankind flew everywhere way before Christopher Columbus fooled Queen Isabella into giving him money to find the land talked about by the northern Europeans. It was a land with huge reserves of copper and who knew what else. Old Chris told Isabella that he was going to China. He and other that came after found Oriental Olmecs, Phoenician Aztecs, and Egyptian Maya. Forget all this about Christopher Columbus and let me get back to Mars. I'm not going to get into this war much at all, because it may not have been significant to our Martian story. Certainly, it was of no concern to Venus. Venus has already been turned into a fireball. Again, let's look at what the book of Jasher tells us.

"Jasher"

Jasher 11:2-7- *And the sons of Noah began to war on each other, to take captive and to slay each other, and to shed the blood of men <u>on the earth</u>, and to eat blood, and to build strong cities, and walls, and towers, and individuals began to exalt themselves above the nation, and to found the* *beginnings of kingdoms, and to go to war people against people, and nation against nation, and city against city, and all began to do evil, and to acquire arms, and to teach their sons war, and they began to capture cities, and to sell male and female slaves-- and they began to make graven images and unclean simulacra, and malignant spirits assisted and seduced them into committing transgression and uncleanness. And the prince Mastema [one of the Anak people] exerted himself to do all this, and he sent forth other spirits, those which were put under his hand, to do all manner of wrong and sin, and all manner of transgression, to corrupt and destroy, -for every one turned to do all manner of sin and transgression--.* War was just beginning, but it would be horrible. Later Jasher indicates that 1/3 of all the people of the world were killed in the war. The Genesis discussion of this war only tells us about the Tower of Babel and it being crushed by 70 angels, but huge piles of evidence around the world paint a picture of destruction that was so bad that people were thrust back into a Stone Age condition and we generally had to start all over. Around the world we find the war affected all people on this planet and we can assume the same type of distress was felt on Mars. Venus was a ball of fire, so any colonization would have had to been to either the Moon and/or Mars.

Flying In India

While hundreds of pieces of evidence convince us that the ancient people flew on a regular basis, The Indians told us plenty. Ancient Indian texts, some written over 2000 years ago, were transcribed from even more ancient texts and traditions. These works provide us with a fantastic view of the ancient world and, lucky, these writings were not destroyed, as were most of the historical works from around the world. Some of the more famous works that were used in this book include "Bhagavata Purana", "Mahabharata", and "Ramayana", but there are many others. In India there were concerns that fearful people would try to destroy the artifacts or use the information contained in them wrongly. Just as was done in other parts of the world, some of the works were hidden and presumably lost. It was believed, general populations should not know about some things and I'm not sure they were wrong.

The surviving Indian texts are some of the best evidence that we have about flying from before the worldwide flood. Even detailed information about how to build them and use them was provided but, unfortunately, the details are somewhat garbled because of the many years of retranslation that must have occurred between the time of the actual knowledge of the technology from before the flood and the time the information was put into these ancient books. Let's look at some of the actual articles. I think you will agree that Flying Machines were high on the list of important things to the people of India even as late as 5 thousand years ago. Stories were told and retold of these ancient wonders. Luckily for us, the descriptions were not lost during the ravages of Libraries that happened around the world. The names of the books are weird, but the information should begin to sound familiar. It should be noted that the reason that we still have information about the ancient flying machines in Indian literature is that no major destruction of ancient libraries was done in this country. The major libraries on the Maya, Egyptians, Babylonians, Chinese and others around the globe, were almost completely destroyed by people trying to protect others from ancient information they assumed to be dangerous, like information about flying machines.

"Brhat-kathd"

This book describes "Dkdsa Vantras" as flying sky machines that were made and used by the Yavanas.

"Ghatotrachabadma"

I really love the names of these books. No! I can't pronounce them, but this one might be the English sentence "Got to track a bad ma". Anyway, like the others, this one is filled with aircraft. Here is a sample. *"The Rakshasa once more came down to Earth in his golden vimana. When it had landed it looked like a beautifully shaped mound of antimony on the surface of the ground."*

"Ayodhya Kandam"

This one has a very descriptive section that gives us great insight into what these things looked like. It stated that the *splendid chariot, made of silver and coated with tiger-skin was as bright as fire itself. It made a noise like the roaring of the clouds. It defied all obstacles. It was adorned with jewels and gold, and was dazzling to the eye. It was bright and swift making a sound like the muttering cloud in the sky. He exited his abode that was like the beautiful moon passing through a huge cloud.* [If you hear a "muttering cloud", look up and you might see a vimana.]

"Valmiki"

The sky chariot which was a wonderful power and wings for speed is gilded and lustrous throughout. It can leap above the hills and valley. [This particular flying machine appears to be more like an airplane with wings, but most were saucer shaped or cigar shaped.]

"Mahavira of Bhavabhuti"

*"The Sky is full of stupendous flying machines [puspaka]. They convey **people** to Ayodya and light up the night sky with a yellowish glow."* [The yellowish glow sounds like mercury compound. We are only now beginning to understand the remarkable fuel mercury really is.]

"Ramayana"

The Ramayana is a set of 7 Books of poetry consisting of 24,000 Couplets written around 300 BC. Like many ancient Indian Historical records, this set of books details information about flying machines. Below are a few of the many references of Puspaka or flying machines.

Puspaka Cars *resembled bright clouds. They rose up into the atmosphere.*

When morning dawned, *Rama, taking the Celestial Car "Puspaka", sent to him by Vivpishand, stood ready to depart. The car was Self-propelled. It was large and finely painted. It had two stories and many chambers with* *windows, and was draped with flags and banners. It made a melodious sound as it coursed along its airy way.*

The Puspaka Car*, that resembles the sun and belongs to my brother, was brought by the powerful Ravan. That aerial car, going everywhere at will, is ready. That car, resembling a bright cloud in the sky, is already in the city of Lanka.*

Vedic Texts

The "Vedas" consist of 4 books written in Sanskrit. The oldest is the "Rig Veda" and is considered to be the oldest actual book, or codex still existing in the world. The books contain magical formulas, descriptions of deities, and history of creation. The set of books were last transcribed in 1500BC. Various types of flying ships were described in these books. Some were called ahnihotra-vimana or elephant-vimana and others had different names. Each type had a different number of engines, and a yellowish-white fuel with a mercury compound was used to power them. One of the Vedas is called "Samaranga Sutradhara". This text devotes 230 verses, to the use of the flying machines in peace and war. Below are some of the details that were given.

Strong and durable *must the body of the Vimana be made, like a great flying bird of light material.*

Inside one *must put the mercury engine. Its iron heating apparatus must be underneath.*

By means of the power latent in the mercury *which sets the driving whirlwind in motion,*

A man sitting inside *may travel a great distance in the sky.*

The movements of the Vimana are such that it can vertically ascend, vertically descend, move slanting forwards and backwards.

With the help of the machines human beings can fly in the air and heavenly beings can come down to earth.

"Puranic Texts"

This Sanskrit book, written in 1000BC, references flying and discusses the requirement to write down the knowledge of the ancients and secure them in a safe place before the flood. The Puranas are sort of like a Hindu Bible. This Hindu Bible is made up of 7 ancient books or Puranas. The details echo our previous information.

"Bhagavata Purana"

Visvakarma, the architect among the Gods, build aerial vehicles for the gods.

Oh, you Uparicara Vasu, the spacious flying machine will come to you- and you alone, of all mortals, seated on the vehicle will look like a deity.

Having made his vow, the foolish King proceeded to worship Salva as his deity - at the end of a year he gratified Salva, who had approached him for protection, by offering him a choice of gifts. Salva chose a vimana that could not be destroyed by Devas, Asuras, humans, Gandharvas, Uragas nor Rakshasas, that could travel anywhere he wished to go, and that would terrify the Varishnis.

Lord Siva said, "So be it." On his order, Maya Danava, who conquers his enemy's cities, constructed a flying vehicle made of iron named Saubha, and presented it to Salva. This unassailable vehicle was filled with darkness and could go anywhere.

Upon obtaining it, Salva, remembering the Varishnis' enmity toward him, went to Dvaraka. Salva besieged the city with a large army . . . *From his excellent vimana he threw down a torrent of projectiles, including stones, tree trunks, thunderbolts, snakes and hailstones. A fierce whirlwind arose and covered everything in thick dust*. The vimana possessed by Salva was very deceptive.

It was so out of the ordinary that sometimes many vimanas would appear to be in the sky, but at other times none.

Sometimes the vimana was visible, sometimes invisible. And the warriors of the Yadu Dynasty were totally confused about the location of this mystifying vehicle: often times they would see the vimana on the ground, sometimes flying in the sky, other times resting on the crest of a hill, and even floating on the water.

That awesome vimana flew in the sky looking like a whirling firebrand--it was never still, even for a moment."

Having spoken thus, Maharaja Nirga made a complete circle around Lord Krishna and touched his crown to the Lord's feet. Granted permission to depart, King Nirga then boarded a wonderful celestial car as all the people looked on.

While Dhruva Maharaja was passing through space, he saw, in succession, all the planets of the solar system, and on the path, he saw all the demigods in their vimanas showering flowers upon him like rain."

He traveled in that way through the various planets, as the air passes freely in every direction. Coursing through the air in that grand and splendid vimana, which could fly at will, he surpassed even the Devas.

"Siva Purana"

*Then the highly intelligent Asura Maya built the cities. There were many palaces with gems. **Aerial cars** shining like the sun, set with Padmaraga stones, moving in all directions and looking like moonbeams, illuminated the cities.*

"Vaimanika Sastra"

Another ancient work that gives us great insights into the ancient flying machines is the "Vaimanika Sastra" The Sastra was, almost certainly, a rewritten document established from very ancient information of the day. In this 1400-year-old work written by Bharadvajy the Wise, 8 chapters and diagrams of three types of flying machines were provided including apparatuses that could neither catch on fire nor break. The book also included information on the steering, precautions for long flights, protection of the airships from storms and lightning. Another interesting section deals with how to switch the drive to "solar energy" from a free energy source which sounds like "anti-gravity." With the "anti-gravity thing, the vimanas could hover in air. The book even described how widespread these vimanas were as it listed 70 authorities and 10 human experts of air travel. Below are just some of the details.

In the 8 chapters, 31 parts of the vehicles and 16 materials that absorbed light & heat were described. Because of the light absorption, they were considered suitable for the construction of Vimanas.

Mantrika: The invoking of mantras which permitted anyone to achieve certain spiritual and hypnotic powers so that he can construct airplanes which cannot be destroyed.

Tantrika: By acquiring some of the Tantric powers, one could endow his aircraft with those same powers.

Goodha: This secret permits the pilot to make his *vimana* invisible to his enemies. A*drishya* accomplishes the same purpose by attracting something called 'the force of the ethereal flow in the sky'.

Paroksha: This helpful hint enables the pilot to, sort of, paralyze other *vimanas* and put them out of action.

Aparoksha: One could employ this ability to project a beam of light in front of his craft to light his way. **[Sounds like a VISIBLE LASER]**

Viroopa Karana: With this skill mastered, the pilot could produce 'the 32nd kind of smoke?', charge it with 'the light of the heat waves in the sky' and transform his craft into a 'very fierce and terrifying shape' guaranteed to frighten an onlooker. **[I have no idea how nasty the other 31 types of smoke were, but this one was really bad.]**

Roopaanara: This can cause the *vimana* to assume such shapes of a lion, tiger, rhinoceros, serpent, or even a mountain to confuse observers.

Suroopa: If one could attract the thirteen kinds of 'Karaka force', one could make the *vimana* appear to be 'a heavenly damsel bedecked with flowers and jewels'. **[I'm not sure how a projection of a pretty girl helped you win a war,**

but the Suroopa stuff was evidently one of the "defense weapons"]

Pralaya: This deadly secret pushed electrical force through the 'five-limbed aerial tube' so that the pilot could 'destroy everything as in a cataclysm'.

Vimukna: This sends a poison powder through the air to produce 'wholesale insensibility and coma'. **[Mean weapon like Mustered Gas]**

Taara: This ability, once mastered, provides the pilot another means of avoiding contact with an enemy or hiding his purpose from observers: 'By mixing with ethereal force just right 10 parts of air force, 7 parts of water force, and 16 parts of solar glow, one could and project it, by means of the star-faced mirror, through the frontal tube of the *vimana*. The appearance of a star-spangled sky is created. **[Whatever that was]**

Saarpa-Gamana: This secret enabled the pilot to attract the forces of air, join them with solar rays, and pass the mixture through the center of the craft so the *vimana* could 'have a zig-zagging motion like a serpent'.

Roopaakarshana This capability permitted the pilot to see inside an enemy's airplane.

Kriyaagrahana: This allowed one to spy on 'all the activities going on down below on the ground'.

Jalada roopa: This would instruct the pilot in the correct proportions of certain chemicals which would envelop the *vimana* and give it the appearance of a cloud."

Aavartaas **or aerial whirlpools:** These were innumerable in the above regions. Of them the whirlpools in the routes of the *vimanas* are five. In the **Rekhapathha** there occurs the whirlpool of winds. In **Kakshya-pathha** there occurs *Kiranavarta* or whirlpool from solar rays. In **Shaktipathha** there occurs *Shytyaavarta* or whirlpool of cold currents. And in **Kendrapathha** there occurs *gharshanavartaor* whirlpool by collision. Such whirlpools are destructive of *vimanas*, and have to be guarded against. "The pilot should know these five sources of danger, and learn to steer clear of them to safety.

"Mahabharata"

This book was last transcribed in 1500 BC. It probably was re-written from works originally made about 5000 BC. Written in Sanskrit it is the longest poem ever written. This gargantuan poem is made up of 90,000 couplets from various poets. The "Drona Parva" poem, detailed below, describes wonders of the ancient flying machines and wars.

The Vimana [Flying ship] characteristics were described in detail. Here are some general ones. *The ships were-12 cubits in circumference, had 4 wheels, rose in air, and as they flew, a charge of mercury caused roaring flames to shoot out.* [**Remember the Mercury from before.**]

The vimana had all necessary equipment. *It could not be conquered by the gods or demons. And it radiated light and reverberated with a deep rumbling sound. Its beauty captivated the minds of all who beheld it. Visvakarma, the lord of its design and construction, had created it by the power of his austerities, and its outline, like that of the sun, could not be easily delineated.*

And he also gave unto Arjuna a car *furnished with celestial weapons whose banner bore a large appreciation and its* *splendor, like that of the Sun, was so great that no one could gaze at it. It was the very car riding upon which the lord Soma had vanquished the Danavas. Resplendent with beauty, it looked like an evening cloud reflecting the effulgence of the setting Sun.*

***Bhima flew along in his car**, resplendent as the sun and loud as thunder . . . The flying chariot shone like a flame in the night sky of summer . . . it swept by like a comet . . . It was if two suns were shining. Then the chariot rose up and all the heavens brightened.*

***And on this sunlike, divine, wonderful chariot** the wise disciple of Kuru flew joyously upward. When becoming invisible to the mortals who walk the earth, he saw wondrous airborne chariots by the thousands.*

***And the celebrated Arjuna,** having passed through successive regions of the heavens, at last beheld the city of Indra. And there he beheld celestial cars by thousands stationed in their respective places [an airport?] and capable of going everywhere at will, and he saw tens of thousands of such cars moving in every direction.*

***Having vanquished his foe**, Krishna furnished with weapons and unwounded and accompanied by the kings, came out of Girivraja riding on that celestial car. Upon that car Krishna now came out of the hill-fort. Possessed of the splendor of heated gold, and decked with rows of jingling bells, it always slaughtering the foe against whom it was driven. It was the very car riding upon which Indra had slain ninety-nine Asuras of old.*

***Thereupon that best of cars became still more dazzling** with its splendor and was incapable of being looked at by created* *beings, as the midday sun surrounded by a thousand rays . . . And Achyuta, that tiger among men, riding with the two sons of Pandu upon that celestial car . . . coming out of Girivraja, stopped (for some time) on a level plain outside of town.*

Those terrible Rakshasas *had the shape of large mounds stationed in the sky.*

Vimanas, decked and equipped according to rule, *looked like heavenly structures in the sky. As they flew away, they looked like highly beautiful flights of birds.*

The Gods came *in their respective flying vehicles to witness the battle between Kripacarya and Arjuna. Even Indra, the Lord of Heaven, came with a special type of flying vehicle which could transport 33 divine beings.*

He [one of the gods] entered *into the favorite divine palace of Indra and saw thousands of flying vehicles invented by the Gods lying at rest.*

Pakistan Destruction

While we're on the subject of flying, I guess you could tell from the previous chapter that there was some blasting going on. One of the hardest hit, during the last of the 3 major wars, was in the northern portion of India. For this evidence we go to the ancient city of Mohen jo-Daro. Here we find drawings depicting the ruling "goddess" who was so strong she could fight two tigers at the same time. She was shown with a flying ship over her head. [See picture to the right] Some have tried to identify it with something else, but none have made any sense.

 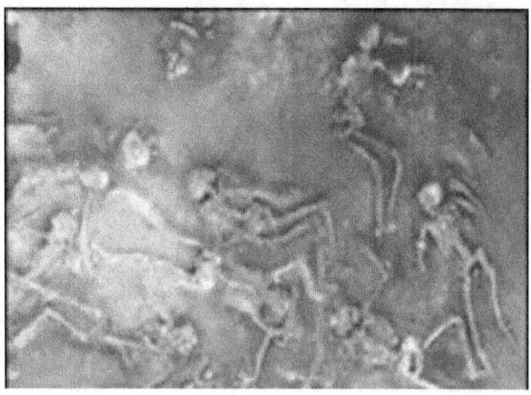

While the written testimony is horrendous, there is also physical evidence is just as bad. While I'm not going to get into most of the physical evidence of war in this book, let me briefly talk about Mohen jo Daro.

The City

Mohen-jo-Daro shows how the flying machines eventually were used. Mohen-jo-Daro means Mound of the Dead. Thousands of people lived and worked in this "modern" city. Pakistan must have been at the center of many trade routes as far back as 7 thousand years ago and, just like the other places, civilization quickly sprang up in the region. Mohen-jo-Daro has been touted as the first city that was built in a very modern way. The streets were all laid out in a grid pattern and there was underground sewage. Throughout the city, widely separated sewage and water lines, were laid out in the same pattern showing the great understanding of sanitation requirements and that the city itself was fully planned before it was built.

All of a sudden, destruction was everywhere. We have no way of knowing how many cities were destroyed, but there have been plenty of battle sites uncovered including cities named, Dwaraka, Betdwaraka, Harappa, and Mohen-jo-Daro. Scientists have already found 140 civilization sites and more are found every year. All the sites were abandoned around the same time period [**around 6 thousand years ago**], so we can get a pretty good picture about why the Indian descriptions of flying machines included elements of their terrible weaponry.

In what remained of Mohen-jo-Daro, hundreds of black lumps of melted clay pots littered the streets and skeletons were found in the street holding hands as if in complete terror during the last minutes of life. As found in other parts of the world [Scotland, France, Peru, United States, Turkey,

Egypt and other places], stones on walls were fused together as if a nuclear explosion occurred. The people had no time to flee. The picture shown below is why the city is called mound of the dead.

Some of the victims we holding hands, some were face down where they fell. Hundreds of bodies were destroyed in an instant as the onslaught of weaponized flying machines gave us more grim evidence of their existence. It was one of the cultural centers of the world during the Tower of Babel days and now it was littered with the dead. The thing that is interesting to me is that the bodies have been said to be some of the most radioactive of all ancient bodies found.

Protection From War

I guess you can tell that the wars got worse and worse. Soon, to protect themselves, guess what the people did. --- Time's Up! The people of the world began to live underground. [I told you that previously. How did you get it wrong?] Certainly, finding underground cities that were used thousands of years ago is an almost impossible job, but vast underground cities have been found around the world. We haven't found most of the underground dwellings, but the ones we have found are very impressive and show great fear about living on the surface of the earth. In Turkey, Malta, and the USA are the best examples, but they certainly aren't the only places. The building of the cities was directly associated with flying. People in aircraft dropped bombs on the people that were on the ground. The ones that survived hid underground.

Mediterranean Underground Cities

The people of that time had to go to extremes to live any sort of "normal" life. They lived under the ground in sort-of bomb shelters. Evidently, they lived in these "homes" for many, many years. No one in their right minds would ever think that an underground city could be fortified enough to keep any ancient military at bay. Simply flooding an area or starving the inhabitants, or sending down fire all would have been easy methods to become victorious. Either the ancient people were all crazy, 6 thousand years ago, or something else was going on. If nuclear weapons were used, any surface dwelling would have been little sanctuary for the people. Missiles fired from flying machines would have not been very successful on underground cities. Laser weapons would not have been effective. In fact, all of the weapons identified in the Mahabharata would have been less successful on underground dwellings than surface ones. Anyway!! Here are a few of the underground towns that have been found to date. I'll let you be the judge and determine on your own whether the people woke up one day and had a terrible urge to start living underground or they were forced into the ground because of some horrible fear that it would be the only way to survive.

Before you decide, remember that the ancient Biblical works of "Jasher" and others indicate that during this time period when the Tower of Babel fell 1/3 of the population died. OK! Now what is you unbiased opinion. --- I knew it! Underground cities help us prove the existence of flying machines during ancient times.

Malta Underground Cities

The drawing shown below is one of the rooms from an underground city on the tiny Island of Malta.

There is no reasonable explanation for their manufacture other than protection from bombs. Although the buried cities were **difficult to defend** with respect to a siege and other common attack methods, they would have been great for keeping flying machines from bombing the city. The underground citadel could house hundreds of people and the workmanship of the underground arena [above] shows clearly that this was not just a bomb shelter that was used for a short period of time. It was home sweet home and constructed with a high level of precision.

In addition to the underground components of the city, the above ground portions of the city clearly show it was ready for massive attacks. One of the main areas of the Malta civilization is shown below. The walls are twenty feet thick in some places and only one story high to limit air strike profile.

Malta is sort of in the middle of the Mediterranean Sea and before the worldwide flood; a civilization had flourished there. Finally, it appears that much of the civilization moved underground. Researchers discovered a society that needed the protection of the underground and built elaborate places to protect themselves and live in style. In Malta, Ghar Dalam is one of several underground complexes. It is a huge complex of tunnels covering over 1600 square miles and descending three stories or as much as 30 meters below the surface. In it, some 7000 skeletons and bones of extinct dwarf rhinoceros were in one area. The inhabitants must have been accustomed to Rhinoceros stew.

Above ground, a high temple measuring 100 x 100 meters is called the Tower of Giants. The remains of huge columns are scattered around the area above the underground haven.

One of the massive columns is 26 feet x 13 feet Diameter. [Now that's a column.] The only thing that is known is that all inhabitants vanished from the area over 5 thousand years ago. The picture shown below is another view of one of the underground areas.

This one is called the Hypogeum. The impressive architecture of this Middle Eastern, subterranean, arena has been determined to be over 5000 years old. Some less than reasonable scientists, somehow, believe that the exact right-angle carvings were done with stone mallets. I say they need to change whatever they are smoking.

Turkey Underground Cities

One of the main "ancient cities" in Turkey has buildings twenty stories high and they can't be seen because the entire city is below the ground. The people dug underground until a massive structure became home for thousands. Imagine how afraid the people must have been to build these deep, multi-level cities.

Cappadocia, Turkey

Below is a drawing of one room in one of the **36 known** and very impressive underground, bomb shelter, cities located in Cappadocia, Turkey. This one could hold as many as 20 thousand people and protect them during these wars. This particular city is actually about 20 stories deep and has huge locking stone doors between areas. Today, only the first 8 stories have been excavated, but they are known to continue on down as much as ¼ mile into the Earth.

Underground Living

These ancient pre-Hittites in Cappadocia were complete settlements. Inside their tunnel cities they had everything they needed including horses, wells, living accommodations, worship areas and many, many rooms. They had all the comforts of home without the worry of bombing. The people lived sort of like ants always worrying about air strikes. Certainly, this way of life would have put the inhabitants in jeopardy of "siege" warfare, having fires lit at the entrances, flooding, and many other nasty ways to be destroyed, but clearly this would have been a very effective way to limit damage from air strikes.

America Underground

Death Valley

For those who think that the Middle East was the only inhabited area of the world at this time, you probably haven't read about the underground cities in the United States. In Death Valley, a series of over 30 caves and tunnels over a 180 square mile area was "home" to a scared civilization. The remains of a civilization were found and reported in the Nevada "Hot Citizen" newspaper. The main reason for the inhabitants moving underground seems to be this ancient war period. In the bomb shelter city of Death Valley, Nevada, there was found the mummified remains of some 9-foot tall inhabitants who evidently wore knee length, leather trousers. There must have been thousands of people living in this place at one time and it really was a complete city. In the city was found a museum type collection of mummified animal and dinosaur remains, a ritual hall, and even gold. All the comforts of home without the worry of bombs and that city were not the only underground marvel in America.

Other Underground Shelters

Around the country, we find underground bomb shelters like the one shown next as people were afraid to come to the surface.

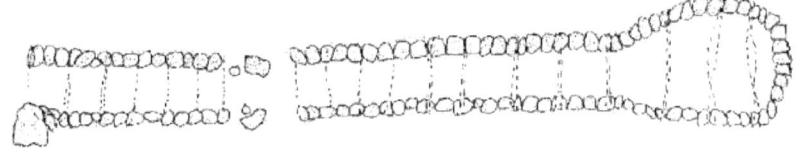

Peruvian Underground City

In Peru there is has been reported to be an underground city with multiple rooms cut into tunnels and even a library and furniture. "Relax and read while the bombs are dropped" might have been their motto.

Popul Vuh Reference

In the "Popul Vuh" of the Maya we read the following, *"Giant tunnels were built by the ancient race of **white** men to protect them from **endless** cataclysms."* [They possibly were built to protect them against the endless air assaults of the wars we are discussing.] The picture below is from a site 200 miles south of Lima.

Guatemalan Underground City

A second group of caverns in South America have also been found and explored for miles. The tunnels were reportedly man made and the full extent of the caverns is still not completely known, according to several researchers including the famous researcher Erik Von Daniken. Ground-penetrating radar has been used recently to determine if tunnels exist under some of the major archeological locations around the world. None of the results of these radar excursions were more staggering than the indications of miles of tunnels in Guatemala that extend from the ancient Mayan city of Tikal to a distance of about 800 kilometers distance from the city. There probably were more people living underground than living above the ground. Underground was the safest place to be.

Middle Eastern Buried Cities

Persian Underground City-For this evidence we turn to historical records. In Persia, ancient writings indicate that ancestors of the Iranian race had escaped a number of long winters of ice and snow by building an underground city. The wars could have caused the desolation and the underground city may have also protected them against bombs. That story is contained in an ancient work called "Vendidad". The Persian history told the story of the Aryan race that existed before the worldwide flood.

The Greek historians at the time of Cyrus placed the writings of the first prophet of the Aryan historian, Zarathushtra, at around 8000 years BC and according to "Vendidad", the sacred book of the Zoroastrians, the ancient Persians had to live underground for a long period of time. Below are excerpts from that work. The underground living points to wars with air assaults.

Excerpts from "Vendidad"

Ahura Mazda, brought Yima [King of the Aryan Nation] a weapon - a "Golden plough" which was dagger shaped with golden forks. [This weapon was a pretty terrible weapon. Evidently is responsible for the Aryans being able to take over huge areas of land during the king's first thousand years of rule. By the way, notice that the long reigns are consistent with the Biblical, Sumerian, Babylonian, and Egyptian histories of before the flood.]

Under Yima's first 300 years of rule, the Aryan land had prospered so much that the land became full of cattle, men, and something called the "red flaming fire". [This was a time of peace, but the "red flaming fire" was probably another type of weapon. Its description is never just fire, but is always depicted as all three words together.]

Yima, expanded his nation to the west using his golden plough. Yima's rule extended another 300 years and he expanded the country to the south and west, making it 2/3rds greater in size. After another 300 years with his "golden plough" he again expanded to the south and west making the Aryan kingdom three times larger than before. [Like I stated before, the war before the worldwide flood was a LONG WAR.]

All of a sudden there was a terrible winter. This time the snow did not melt. The corporeal world was somehow DAMAGED. The whole land was full of hard Ice. [The earth shifted during the preflood wars. In this area, the land went closer to the poles and froze. It is believed that this was about 12 thousand years ago.]

To protect themselves during this terrible time, they built and entered into a VARA [city]. Inside they could not see the stars, because the Vara was underground. The Vara had its own artificial lighting. [This artificial lighting is an example of the technology from before the flood, but notice that, like other nations of the world, the Aryans also moved underground during these terrible wars.]

Iraq Underground-In Iraq the same type of underground dwellings was found. There is little doubt that around the world, no one wanted to be on the surface during the wars. The picture next left shows one site in Iraq.

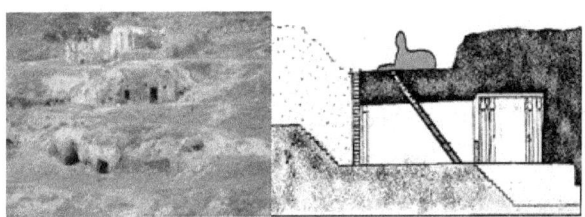

Egyptian Underground

In Egypt, a huge labyrinth of tunnels was found using SIRA radar. The tunnels under the Giza surface have been studied for years, but only recently has one of the key scientists named Dr. Hurtak shown film footage of the underground multistory metropolis. He estimated that the structures were about 15 thousand years old and the videos reportedly include huge cathedral-like structures under the Sphinx area as shown above right. I expect we will hear much more of these findings in the near future. They may have been living underground in fear of the wars outside, but the Egyptians did it in style. From the sphinx, a huge complex of tunnels allowed people to go to the various pyramidal structures in the Giza Plain as shown below.

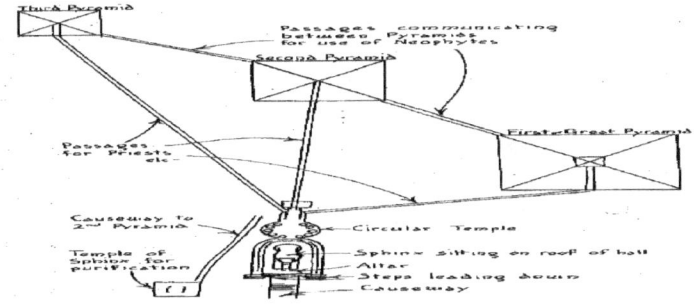

Underground In China

The fear of war from the sky in China is evident by what we have not found. Soon we may know about how people had to live underground in China during the war, but right now we have pipes and a pyramid. They were found about 40 kilometers to the southwest of Delingha City in the western province of Qinghai on top of Mount Baigong.

Some people found a pyramid next to a lake way away from any civilization, but that isn't the thing we need to be concerned with here, because they also found pipes. Next to the pyramid are the remains of three caves. Two have collapsed and a third is still open as shown in the picture below. The upright lines in the center of the cave is actually a pipe going under the ground to a place that has not been excavated as of yet. This iron pipe and the other pipes found in the area are clogged, but once they must have been used to go from somewhere below the earth to the surface.

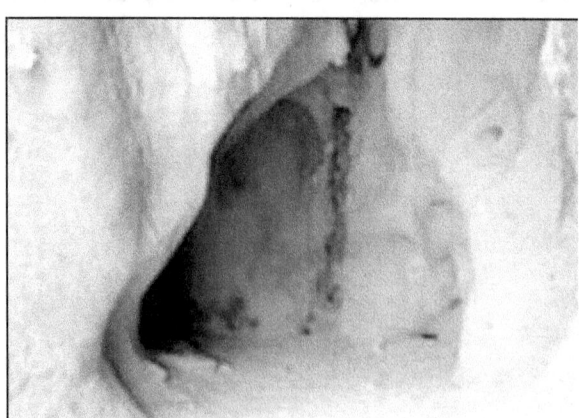

Another underground village was found in Gaochang, China. The part that is visible on the surface is shown below.

This underground living is found around the world and probably occurred around the time of the Aryan wars, as the iron pipes have not yet completely decomposed. It is estimated that there are over 30 pipes of various sizes in the arca with some actually in the adjacent lake, some visible on the ground and about a dozen in the caves. The pipe material has been tested and the result shows that they are made up of 30 per cent ferric oxide with a large amount of silicon dioxide and calcium oxide. The large content of silicon dioxide and calcium oxide is a result of long interaction between iron and sandstone, which means the pipes are very old. The investigators also indicated that eight per cent of the content could not be identified. We may be able to learn more about our predecessors when we excavate the area, but so far all we have are pipes.

I know you think that I am spending too much time on living conditions of humans rather than how the wars escalated enough to make colonization of other planets reasonable, but I think it is important to understand flying and the art of war had developed to such a high level these many thousands of years ago that underground living became ALMOST the only reasonable way of survival. The only other option to gain peaceful life was to venture outside of the earth. Certainly, the living conditions weren't the best, but they weren't so great on the earth at that time either. Real relief was found by leaving the planet in a flying ship.

What About the Other Wars?

I'll bet you are wondering why I am spending so much time on this war that happened only 6 thousand years ago and worse wars happened 120 thousand and 15 thousand years ago. The reason is that the other wars were so long ago much of the evidence has been lost. This one is still new. As each war began, we can believe people left their homes and traveled to the other available sites. By this time Venus was not a good place so Mars would have found some eager colonists who would, again, have to live underground to support manufacture of air and food stuffs. Later we will see that there have been attempts to terraform Mars with some successes.

Flying In Space

Like I said, these flying people went into space and what better place to go than Mars. Many of these accounts appear to be depictions about going into space after the worldwide flood of Noah [10000 years ago]. Possibly many occurred during the massive destruction of the war 6000 years ago.

If we read in the Book of Enoch, we find even a clearer description of humans visiting other planets. *Enoch 74:15-I saw likewise the chariots of heaven, running in the world above to the gates in which the stars turn, which never set. One of these is greater than all which goes around the world.* [The writer of the book of Enoch not only described planetary travel, but also the existence of a large space traveling vehicle that orbited the Earth.]

The Cabbala And Space

The ancient Jewish Cabbala texts provide detailed information about seven of the inhabited planets that were known at that time. Here is what the Cabbala has to say about these planets. Where the planets actually are is not known, but the ideas that some have 2 suns, some have dim sunlight, or ancient man would not naturally know about the possibility of a red sun, as well as other things. This is just another piece of data to help us determine the truth. It is looking like planets were inhabited and some of the inhabitants came to earth. One reason to come here would be that the "people" were from here originally. As you read through these descriptions, you are free to wonder how they could possibly have known about these non-earthlike abodes. The idea that the descriptions were fabricated is improbable simply because of the bizarre descriptions presented.

Geh-The inhabitants sow and plant trees. They don't eat wheat or cereal. They eat from the trees and their world is shadowy There are many large animals there.

*Nesziah-The people there eat shrubs, but the food grows without having to be planted. The people are small and have **only small holes instead of noses**. They are also very forgetful and when they do work, they often do not know why they started it. They see a **red sun**. [**Maybe the people that have been seen coming in UFOs came from here.**]*

*Tziah- The inhabitants of Tziah must not eat what other beings eat. They constantly seek out underground watercourses. These people are very tall and handsome and worship God more than other races. The planet has great riches and handsome buildings. The ground is dry and they have **two suns**.*

*Thebel-The Thebelese eat only from the water and are superior to all other beings. On Thebel, all five races live. The world is divided into zones and several variations in skin color can be found. They can make their dead come back to life and the **world is far away from the sun.***

Adamah- The inhabitants of Adamah are also descended from Adam because Adam complained about the cheerlessness of Erez. They cultivate the earth and eat plants and animals. They are also sad people. In the past they had been visited by Thebel, but the visitor Thebelians had been struck by failing memory and don't remember whence they came.

Erez-*The people of Erez were, descendants of the "first man" that lived on our planet. They cultivate plants and animals. They are sad-faced people who often make war on each other. There are days in this world, and the groupings of constellations are visible.*

Arqua-*On Arqua, the inhabitants are mostly farmers. Seasons are long, sowing and harvest recommence after several years. They can speak many languages and visit many of the worlds. Their face is different from ours.*

Please don't be taken aback by the bizarreness of the accounts as this detail would have been rewritten over hundreds of years and each time it would change a little.

More Space Flights

All over the Middle Eastern world the descriptions and evidence of space venturing flying machines is overwhelming. Here are several more examples. The first is an entire ancient book called "Sifrala". It was some kind of technical manual for the flying vehicles. According to this unusual Sumerian tablet, being a pilot was a great honor back in the good old days.

Sumerian Space Flights

Besides the written technical description space flight was described in other historical writings. There can be little doubt that space-venturing vehicles were commonly known about by the ancient Sumerians. Here is one example of their descriptions of a spacecraft. It explains how the rulers of Sumeria flew in them. In this case the queen was considered to be a god. Her name was Ishtar.

From the Sumerian "Rape of Ishtar" story- [We find long distance flight in one day]. One day my queen, after crossing the heavens, crossing Earth, after crossing Elam and Shubur, after crossing… the hierodule [flying ship] approached weary, and fell asleep. Not only did the Sumerians leave behind written text, there is also physical evidence.

Babylonian

In ancient Sumeria and Babylon, many, many depictions of flying rockets, space ships were found. These drawings are some of the many depictions. Apparently, the gods used these vehicles, and carvings were made of what the people saw. The objects that the ancients saw were drawn everywhere. PLATE 4D shows a couple.

Sumerian

This drawing below right is of a Sumerian Model that was hidden away in a museum. It looks like our space shuttle, doesn't it, even down to the 3 blast engines. The model was thought to have been a very old fake for many years, however, renown scientist and linguist, Zecharia Sitchin, examined the artifact in detail and found that it was made from a very light volcanic stone, that it was manufactured in an ancient time, certainly well before ANY thought of a space shuttle could have been imagined. It was determined that, in all likelihood, the Sumerians as indicated by an original geological document manufactured it. To the left are a couple of the other depictions of flying ships.

Pilots from Nepal

In Nepal, an indication of a strange visitors was found. A carved plate has been found which was manufactured around 4 thousand years ago. The carvings on the plate shows a large headed being similar to that depicted as one of the UFO pilots currently seen an elliptical shaped object above him, which is very similar to reported UFOs of today.

That is odd enough, but what I want you to see here is that it looks like the Big-headed guy is traveling from Mars down to Earth in this football shaped thing. There also is a monkey looking thing coming from where Venus would be and a Lizard thing possibly leaving Mars, but let's mostly look at the Martian.

 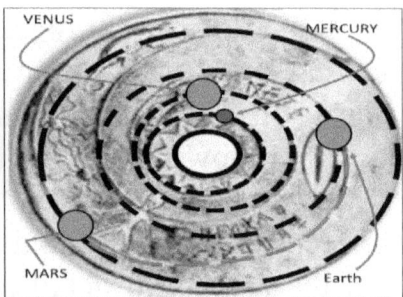

Egyptian Space Flights

The excerpts below are from the ancient Egypt. They are from a number of sources, but they all say the same thing. People knew about space travel.

"Emerald Tablets"

While the details were supposedly transcribed from the teachings of one of the gods named Thoth, the description of flying is unmistakable. *"Everything is good but only to be replaced by the spawn of a lower star --- we entered the great ship. Upward we rose ---suddenly over it rose the waters, vanished from the Earth."*

Pyramid Texts

A section of the "pyramid Texts" identifies the Pyramids as *"ramps to the Sky"* so that *"man can go up to the heavens"*. **[Possibly referencing landing beacons]**

Egyptian History

During Zep Tepi [first time], flying gods came down to earth, flew through the air in flying boats, and gave man wisdom.

Gnostic Space Flights

It is very important for us to examine the Egyptian Gnostic records, because they tried to interpret ancient texts rather than just reporting on it. They wrote what they believed the more ancient texts were TRYING to say. Because of their zeal, the Gnostics interpretation disrupted the true meaning of the words, but in other ways it allowed us to get a different insight into our history. Like the others, these excerpts talk about space travel.

"Origin of the World" Space Flight

This excerpt is from one of the ancient texts found in Egypt. It is typically called "On the Origin of the World". It generally confirms the flying history being presented. Like all Gnostic writings, it is written in a mystic style that typically limits understanding and frustrates investigators, but don't let its mysticism keep you from gaining the information.

The troublemaker that was below them all destroyed the heaven and his Earth. And the six heavens shook violently; for the forces of chaos knew who it was that had destroyed the heaven that was below them. And when Pistis knew about the breakage resulting from the disturbance, she sent forth her breath and bound him and cast him down into Tartaros and when they had become disturbed, they made a great war in the seven heavens. Then when Pistis Sophia had seen the war, she dispatched seven archangels to Sabaoth from her light. They snatched him up to the seventh heaven. [A possible interpretation-During the first heaven wars, the seven archangels fought wars all the way to the 7th heaven. This implies flying between heavens, but there was even a description of one of the vehicles. Several other ancient texts tell us there were at least 10 heavens, 10 livable planets, and seven archangels. The wars must have been huge to affect these planets.]

And before his mansion he created a throne*, which was huge and was upon a four-faced chariot called "Cherubim". Now the Cherubim has eight shapes per each of the four corners, lion forms and calf forms and human forms and eagle forms, so that all the forms amount to sixty-four forms* [The Flying machines were similar to the flying machines seen by Ezekiel. By this, they were used in the "Heavens" War.]

Then Justice created Paradise*, being beautiful and being outside the orbit of the moon and the orbit of the sun in the Land of Wantonness, in the East in the midst of the stones.* [The stones are the same stones indicated in the Bible, or uninhabitable Planets. It, therefore, indicates that the heavens are beyond the planets known as stones.]

"Revelation of Adam"

This section can be found in the ancient Jewish work "Revelation of Adam to Seth" which was also part of the Nag Hammadi texts found in Egypt. Some indicate that this is a revelation of our future, while others insist that it is our past. This writer does not know it, but if it was our past, the people flew into space.

It was then that the Archeon [beings that fought against God in heaven] of the Occident and the Orient set to work to educate mankind in his malignity and to undo all the teachings and counsels of the Wisdom of Truth—but the Archeon failed. [These "bad" Archeon humans passed perversity to man. From other texts, these included genetic manipulation, becoming witches, performing alchemy, and something to do with eating blood. I don't know what the Orient had to do with anything and I'm not sure what the blood thing was all about. It doesn't appear that God cared if a little blood was in meat that was cooked, so they were talking about huge amounts of blood. It makes you think about vampirism, doesn't It?]

Then opened a new period, which altered the circumstances. Then the Archeons sent to this man their Counterfeiter—They were all looking to him, the perverse one, to perform a prodigy. He reigned over the whole earth and over all the sky. [The implication is that one of these Archeon humans controlled the whole world and possibly the "near Planets". For this to happen, flight must have been common. If you are wondering who would have been called the counterfeiter, think about the guy named Satan.]

Greek Space Flight

The Greek histories are more easily interpreted. We find flying in space evident in both Greek history and their "almost history" typically called Mythology.

"Theogony" Space Flights

The history entitled "Theogony" clearly described flying. In Hesiod's book, the Greek version of the history during the early time [before the Tower of Babel time] can be easily distinguished. This section contains descriptions of flight.

A heavy quaking reached dim Tartarus [possibly a planet that was dim in the night sky] and the deep sound of their [Titan's] feet in the fearful onset and of their hard missiles. So, then, they launched their grievous shafts upon one another--And the cry of both **armies as their shouts reached the starry heaven;** *And they met together with a great battle cry.* [My interpretation would be the giant humans called Titans had a Civil War between their various groups. The war escalated until it included battling on other planets, which would have required flying.]

Zeus was filled with fury and he showed forth all his strength. From Heaven and from Olympus he came forthwith, Hurling his lightning. The bolt flew thick and fast from his strong hand together with thunder and lightning, whirling an awesome flame. [Lightning Weapons were used in the wars. We will look at this in more detail later, so don't just think it's a silly notion.]

The life-giving Earth crashed around in burning, *And the vast wood crackled loud with fire all about. All the land seethed, and the hot vapor lapped round the Earthborn Titans [**"Earthborn" indicates they were human**]: Flame unspeakable rose to the bright upper air [**Outer Space**]: The flashing glare of the thunderstone and lightning blinded their eyes for all that there were strong. Astounding heat seized Chaos:* [Finally the wars included something similar to Nuclear weapons and battles extended into outer space with flying machines.]

They launched from their strong hands and overshadowed the Titans with their missiles, and **buried them beneath the wide-pathed earth**, *and bound them in bitter chains When they had conquered them by their strength for all their great spirit, As far beneath the earth to Tartarus.* [Some may recognize the similarity of the Titans being chained and placed in Tartarus and the Biblical version with the Anak being chained and sent to Tartarus or hell. In the Greek version Tartarus was beneath the ground, the Jewish version is less specific, but there is an indication that the bad place is "Below".]

Greek Mythology Space Flights

The better-known mythology has a similar observation that men knew about and went into space. According to mythology, Orpheus, Apollo's son, said the following:

*Those innumerable souls, **they fall from planet** to planet and, in the abyss of space, lament the home they have forgotten.*

This segment could have been talking about the wandering angels and their lamentations about losing the heaven wars or it could be referencing human space travelers. The main thing to see here is that traveling from planet to planet was believed to be reasonable and that living on various planets was considered reasonable to the Greeks just like everyone else in the extremely ancient past. Many of the ancient people knew about major elements of our solar system.

Northern European Space Flights

I know I haven't described the details of the Dropa yet, but just trust me a little longer. These Dropa people had crash-landed in China some 12 thousand years ago, but they were much too far away for this area to have had any contact. Someone told the Norse about various livable planetoids and it wasn't them. The extremely old traditions of the area were collected in a book called "Edda Elder" around 1220 AD. This book tells about many conflicts between God and giants and it also tells about 8 different worlds; each with a different type of inhabitant. While these could be levels of heaven, it appears that these were descriptions of actual planets. Who knows where they got the information for this book?

Asgard [Moon]- *The Gods resided there. Thor, one of the gods, came down from Asgard and protected man from the giants with a mighty weapon similar to a hammer. His adversary was the serpent, Loki, who was the evil god [handsome and witty]. Loki was the originator of all lies.* [As strange as it may seem now, this could well be talking about the moon. This would imply that someone lived on the Moon for some time. Later we will look more closely at the moon and show how this was possible.]

Milgrad [Venus]-*Men lived on Milgrad. In the early days people could travel from Milgrad to Asgard.* [It is believed that this planet was none other than Venus. While the planet is not habitable now, there is strong evidence that its present stated was initiated by a terrible cataclysm that occurred only 12 thousand years ago. Before that time, Venus, most likely, supported colonists.]

Jotunheim [Mars]-*Great giants lived on Jotunheim. In early days giants came to Milgrad from Jotunheim.* [This would have been some place with low gravity or giants would have had trouble. The most likely place was Mars. If the Martians went to the moon, coming back to Earth would have been the next step.]

Vanaheim [Europa?]-The Vanir lived on Vanaheim. [No other information was presented. The next possible inhabitable place would have been one of Jupiter's moons-possibly Europa.]

Alfheim [Callisto?]-*Normal Elves lived on Alfheim.* [The next possible planetoid that could have sustained life would have been Callisto- another of Jupiter's moons.]

Svartalfheim [Granymede]-*The black elves live on Svartalfheim.* [A third moon of Jupiter, Granymede, like the others has indications of previous life.]

Nidavellir [Titon]-*The dwarfs lived on Nidavellir.* [A moon of Saturn has some indication of previous life, but not nearly as much as those orbiting Jupiter.]

Niflheim [Encaladus]-*There is only bitter cold on Niflheim. No one lives there.* [This sounds like the other terrestrial class moon of Saturn. It is almost completely covered in snow and ice, while it looks pretty, living on that planetoid for any period of time would have been practically impossible.]

African Space Flights

Guess what? In Africa we also find space ships. To the Midwest, the Dogon tribe remembered a flying machine. A little farther south we find similar stories from the Luba.

The Luba

The Luba tribe of Zaire had many ancient traditions. Some depicted flying as indicated below. This is talking about humans coming down from the sky or the heavens [in Flying Machines]. In fact, it specifically indicates that in the distant past humans were forced out of the heavens. We will get into that a little later. After the world was created, the most high God and humans lived in the same village in the heavens --They began to quarrel. The humans were expelled to the Earth.

The Dogon

In their tradition, they mentioned that when the Nommo arrived, their flying ship caused a whirling dust storm as it skidded, and that the flame went out as it touched the Earth.

A Planet Revolving Around Sirius-The Bible, the Tanguts, and many other ancient histories all seemed to know about life on other planets, but their descriptions were nothing compared to the Dogon tribe in Africa. From information that has been provided to us from the Dogon, we can be fairly certain that life must have flourished on other planets. The Dogon tribe provided us with a wealth of knowledge about the inhabitants of a planet that revolves around a star we call Sirius. Supposedly, their gods [visitors] came from this planet. The primitive Dogon could not possibly have known some of the information they obtained from these visitors. This wealth of knowledge included the following:

• The concept of dwarf stars,

• The idea of a planet revolving around dual suns,

• The Knowledge of the existence of unseen remote planets,

• And other astrophysical phenomena that have, only recently, been verified

There is also another reason why we should believe that the Dogon outer-space visitors were real. It is the way that the Dogon depicted these humanoid creatures. According to the Dogon, these visitors were not the beautiful angels depicted in paintings. They were ugly humanoid creatures.

Ugly Dogon Gods-According to the Dogon, their "gods" were ugly reptilian or amphibian-like creatures. Naturally, this does not go in line with reasonable thinking about a "God" image. To make this concept even more bizarre, their reptilian god concept was also recognized by ancient Sumerians. Both societies essentially indicated that one of their gods was nasty looking. Try going up to the most powerful person you know and tell him he is ugly and see if he will be your friend. Unless he was truly nasty looking, you might be finding out how powerful he really is. Even then, it's a gutsy thing to do.

Dogon Planet Knowledge-One of the ways we know that these reptilian humanoids actually came from space and visited the Dogon is that the primitive Dogon, without the use of a telescope, were convinced that there was a dwarf star companion to the star Sirius. Even the concept is strange. We now know they were right.

The Tangut Tribe

This central Asian tribe had a strange concept that went along with the apparent Biblical concept of inhabited planets. They believed there were 11 major luminaries including the sun, moon, Mercury, Venus, Mars, Jupiter, Saturn, Tsi-Tsi, Ouebo, Rahu, and Ketu. How could they have known about other planets beyond Saturn? No one knows the answer, but they knew about them just the same. This raises an interesting question that doesn't stop with the Tanguts. How did the ancients know about planets that were too small to see with the naked eye? [The way they probably knew was that some of their ancestors had visited the places.]

Indian Space Flights

"Ramayana" Evidence

The "Ramayana" said the following. *It was a self-sustaining flying city that **traveled in outer space.** One of these cities was named Hiranyapura (city of gold)*

***Beholding the Puspaka car** coming by the force of will that Rama, the king got in, and the excellent car, at the command of Raghira, rose up into the **upper atmosphere.** In that car, coursing at will, Rama was greatly delighted.*

The Ramayana, also contains a highly detailed story in it of a trip to the moon in a Vimana, and in fact it also details **a battle on the moon** with an "Asvin" airship.

"Amsu Bodhini" Evidence

Information about planets was found in the book "Amsu Bodhini" which was not decoded until 1931. It contained information about different kinds of light, heat, color, electromagnetic fields, solar energy, capability to send messages by cable, and **machines to carry people to other planets.**

Secret Society Evidence

According to ancient records, the "Secret Society of Nine Men" wrote nine books, one on **gravity control** was known to ancient historians but is now lost. None of the other books have been found or seen in recent history, as far as we know.

Palm Leaf Evidence

The Indians had their own manuscripts containing information about other planets. The "Palm leaf Manuscripts", found in Karnataka, India are some of the earliest forms of actual books. They contain details about other planets and different types of light. The estimated date of the books is 1000 BC, well before our modern telescopes were invented.

Far East Space Flights

In China they have found documents written in Sanskrit telling about the capability of air travel similar to the information from India. One example, I briefly described earlier is the large group of disc type tablets that have been found and translated. They talk about a flying ship that crash landed and the occupants that survived for many years. The estimated time of the crash was about 10,000 BC, which was probably during the time of one of the worst wars ever witnessed by mankind. That find and many more, show that flying machines were seen, flown and used for war in the area. Here are some examples of the written testimony.

"Collection of Old Tales" Evidence

In this compiled set of still earlier stories from China, an enormous ship on the sea was described as being able to sail to the moon and the stars. The boat was called "the boat to the moon" and it was seen for 12 years. The stories were from very ancient times, but were compiled 1600 years ago.

Chinese Dragon

Ancient texts told of flying dragons covered in armored scales with eyes that flashed lightning and fiery breath that shriveled towns. [Although dragons were probably amazing creatures, some of the descriptions of dragons sound more like the Indian flying machines than living creatures.]

Maotse Evidence

In an ancient Chinese reference called "Shooikng" we find reference to the Tower of Babel incident of the Bible where people lost the ability to communicate. In this reference we find that the same thing happened to the Maotse. The Maotse lost the capability to communicate but it goes on to indicate that they previously talked to people from the sky. It also indicates that they could no longer go to the sky. Presumably they lost more than the power of communication. They lost the memory of flying. When the Maotse brought trouble to the Earth, Changty [God] saw that his people had lost virtues and halted all communication between sky and Earth.

Tibet Flying

In western China in the Tibetan area, their history and folklore are filled with instances of space travel. The ancient works "Tantyua" and "Kantyua" both talk about flying ships as "pearls of the sky". They also indicate that the knowledge of their secrets was kept from the masses.

Korean Dragon

In Korean mythology, the 5-dragon chariot [Oryongeo] was used **to descend to Earth and ascend to heaven.**

Kunming China Evidence

Engravings of rocket-like ships were shown climbing towards the sky. They were found in a pyramid shaped building that emerged from Lake Kunming after an earthquake caused the water level to drop. By the way, just to keep you from thinking that pyramids in China are absurd, I need to straighten one thing out. **There are more**

pyramids in China than in Egypt and some are bigger than the Great Pyramid. There are over 100 pyramids located in the Sianfu area alone. During the olden days China was a popular place. It possibly was a popular place for people in flying ships.

Dropa Evidence

Here I'm back with the Dropa again with a little more detail. In the Lhasa region of China or Tibet, we find compelling evidence of flying and humans living on other planets. The evidence points us towards the fact that during very ancient times humans colonized some of the near planets. When I talk about ancient times here, I mean tens of thousands of years ago. There is a lot of evidence of the colonization on Mars that we will get into that a little later, but Venus is the more likely place for the colonialists to have come from. On occasions the inhabitants of these colonies ventured back to mother earth. One dramatic and sad piece of evidence to this probability is the story of the Dropa. One might consider this physical proof of the colonies and the terrible incident on Venus that occurred 11 thousand years ago? In the Baian-Kara-Ula Mountains of the Himalayas, a set of 716 stone "disk books" have been found in a cave. Several of these disks are shown next.

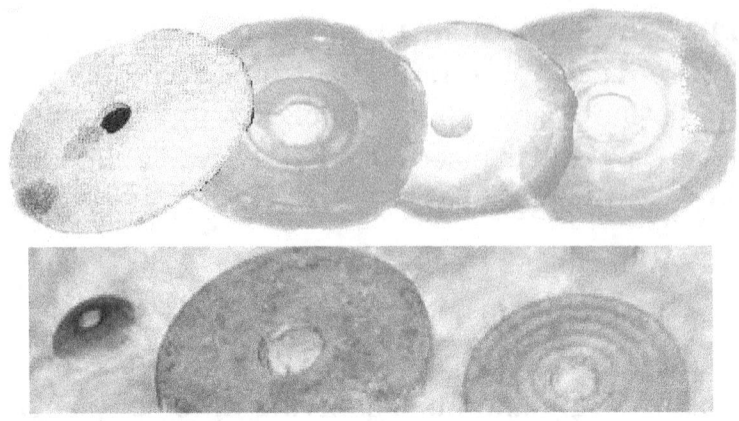

Each disk has tiny hieroglyphs along their surface that spiral outward on the side in one continuous stream. Recently some of the messages they contain were translated. According to the information on the disks a very unusual circumstance occurred and was recorded. NO! They are not money stones from the Island of Yak. As I briefly mentioned before, these "RECORD BOOKS" tell about a family group called Dropa that survived a crash landing of a ship from space. Tibetan traditions provided us with a description of the people. Evidently, they were small, had yellowish skin, and had a large head and large eyes. On the walls of the caves were drawings of the earth, moon, other planets and stars but, according to one researcher, "They were of star patterns that seemed strange." Dotted lines connected the moon, earth, and some of the other indications. [Presumably showing trade routes or some other important track?] That wasn't all. In 1938, researchers found rows and rows of graves. The beings in the graves were originally thought to be some ancient apes. The heads on the tiny skeletons were large and the bodies were spindly. All were less than 5 feet tall. They weren't apes at all. They were small humanoid figures. They were the Dropa.

Even more information can be used to substantiate the information. One of the tribes in the area today are miniature. Like their ancestors they call themselves the Dropa. Guess what they look like! They have large head and light blue eyes. These people also have almost no hair and the average height of the individuals is only about 4 feet tall. One of the individuals was reported to be only 2 feet tall. That being said, here is what researchers have pieced together from the artifacts and data.

Dropa Review

Some people called "Dropa" crash landed on this planet and took up residence. The Tibetans were afraid of these small bodied, yellowish skinned, large headed people and killed quite a few before they realized that they meant no harm. The "space" Dropa stayed in the caves of the area and were never able to leave this planet. After a time, the Dropa assimilated into the community. According to the decoded sections of the stone tablets, the Dropa came here about 11 thousand years ago, but that doesn't mean it wasn't common for others to visit Earth at even much earlier dates. The date may represent the last of this particular type of visit, however. In my book "The Day Venus Exploded", I investigated the shocking probability that Venus once had a moon that exploded about 11 thousand years ago. The aftermath of the explosion was the destruction of Venus as an inhabitable planet. There are literally thousands of pieces of evidence that strongly suggests the scenario above and dates the incident precisely to the time that the Dropa became stranded. They may very well have been the last departing family from a doomed planet.

No Venusians

By all indications, those living on Venus were not Venusians, they were humans that had been living on Venus as a matter of colonization. Over the years, their general characteristics stayed constant—2 eyes, 2 legs, etc. The only things that changed were minor features to allow better adaptation to the climate on a particular colonial site. In this instance, over thousands of years, people were smaller, had larger heads, had less hair, and larger eyes. We can believe that colonists that lived on Mars could have had these same characteristics over time at a location with lower gravity and less light.

Pet Peeves

You may be thinking that the similarity of 2 arms, 2 upright legs, a mouth, nose, 2 eyes; upright walking doesn't mean the people were human. I will tell you this does NECESSARILY mean we have at least one common ancestor. Even on our own planet there are huge differences in the structure of animal types. To even suggest that an alien entity would have almost identical characteristics when a man and a snake "evolved" on the SAME planet would have such differences shows how utterly nonsensical the idea that some remote planet would have parallel evolution that would be a carbon copy of earth. The very odd characteristic of only using 2 appendages to walk with is odd to say the least, but the visitors ALL walk on LEGS because they come from EARTH.

No! This common ancestor probably did not colonize earth from some distant star as some have suggested. Instead, the far stronger probability is that humans ventured FROM earth and some are returning. The only reason that people do not believe the colonists came from earth is that the evidence of

"Modern Man" living on earth for millions of years has been suppressed by those who wish to make our history more comfortable than the truth will allow it to be.

Thai Ancient Verbal History

In Thailand we find a story that sounds familiar. The "lords" of the sky were called the THENS. Their story goes something like this--

In the beginning the earth and sky were connected by a rattan bridge. The sky was controlled by the THENS [Sky humans] The THENS commanded the earthbound humans to bring food as a token of respect. The people refused and the THENS, who controlled the elements, brought a flood to drown everyone. Three great humans named Khun K"an, Khun K'et, and Pu Lang Seung built a flying vessel and sailed into the sky to talk to the THENS. The THENS allowed the humans to come up to the sky until the floodwaters subsided.

The Thai people had passed down information about people leaving the planet in flying ships, not because they were great writers of fiction. The reason was that someone actually witnessed the departures.

Pacific Island Space Flights

The ancient people of Australia and New Zealand were aware of all this space flying going on and wrote down indicators of their knowledge.

New Zealand Remarks

Legends indicate that *the Banana bush was brought by the Manu. The Manu were lofty spirits and protectors **from another star. The Manu were much further along in evolution**.*

Australian Carvings

In Australia, the aborigines drew flying ships on the cliffs. Some looked like they had landed, while others were shown over water and even windows are shown. **[See below]** The drawings are believed to be at least 6 thousand years old.

Central American Space Flights

The Aztec, Maya, and Olmec all showed what may have been rocket type flying ships. Also, we have found indications of visits to outer space. Some examples are described below.

Codex Nattal

In the Nattal, an Aztec Historical Pictographic Document, one of the pictorial images clearly shows what is believed to be a rocket ship with thruster engines and flames.

Olmec Rocket

The Olmecs carved the representation of a rocket-like object with the body of one of their gods at its base. This could have been a headdress if the person at the base was very, very strong, but I doubt it.

Pre-Mayan Helmet

If that is not enough evidence, there is the space helmet shown below upper right. Found in Ecuador, this preMayan relic is not some festive headdress worn in a parade, the visor elements and the full covering of the face, the characteristic "crash helmet" design, the external appendages all point to the same thing. This was most likely a helmet used during high speed maneuvering and quite likely a helmet to insure air supply when none is available.

In a later section we will look at more of the space suited people depicted in statuettes and drawings from around the ancient world so don't go thinking this is just some fancy headgear or an ancient motorcyclist.

Around the world we find these "helmets" and begin to wonder what they were used for.

More Space Helmets

Besides the Mayan helmet found in the Americas, the ancient artists of ancient times may have been trying to tell us about people who needed to have a special helmet to breath similar to the helmets our astronauts use.

Australian Helmet covered the head, sometimes with radio antenna sticking out of the helmets. Other times the helmet and suit seem to all be combined, just like our suits of today. Some depictions show the helmets people raising their arms to the stars or flying space craft similar to those seen today. A few of the depictions are shown next.

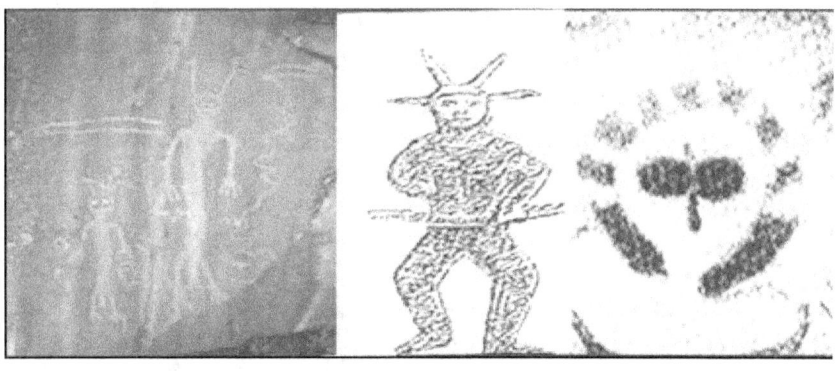

South Pacific

Temehea Tohua is the ancestral home of Queen Vaekehu who was considered to be the last queen the Taiohae. The site of her power contains a number of carved stones brought from around the Pacific The Marquesas Islands were settled by Polynesians around 200 BC and have cultural and language links with other Polynesian peoples across the Pacific. The images below left certainly are helmets and seem to be ready for space travel to Mars or wherever.

 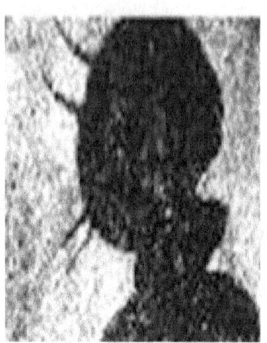

Uzbekistan

In Uzbekistan helmets had radio antenna as shown above right.

Sahara Desert

In the Sahara space helmets also had antenna. Some showed the entire space suit and a flying vehicle overhead. The cave painting [left] is from Tassili, Sahara Desert in North Africa. It dates back to 6000 B.C. The figures do not look human. Notice the flying disk in the sky.

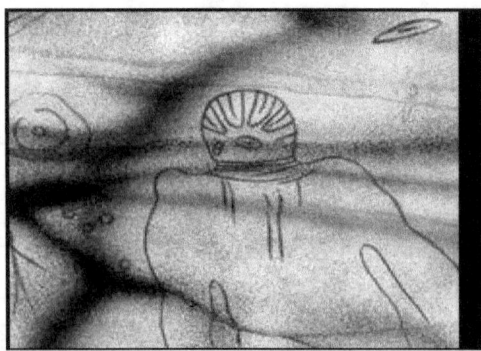 

Italy

Italian helmets had both antennas and a full covering of the head. Here are a couple of the many images found to date.

 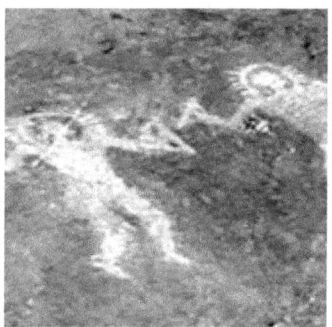

South America

In South America, a small statuette of what appears to be a space suited human. The clothing doesn't look very comfortable or useful in the South American environment. The windshield, bolted on headgear and covered feet look almost like undersea gear, but this was apparently for wearing above the water. Perhaps it was for "<u>well</u> above" the water.

Central America

"Chilam Balam"-Mayan History described in "Chilam Balam" should be looked at a little more closely to show how it describes space travel. "Chilam Balam" states the following:

Creatures arrived from the sky on flying ships. White gods could fly above the spheres and reach the stars.

Behold the divine Maya aloft in a circular chariot of gold, measuring 12 thousand cubits in circumference and able to reach the stars. **[Mars as well]**

More "Popul Vuh" Evidence

"Popul Vuh" states-*Hunahpu, Xbalanque, and Quetzalcoatl* **returned** *to the stars after their life on earth ended.*

Helmets

In Central America, the Maya probably were the most obsessed with space travel and covered the head as well. Strange artifacts were found 55 years ago by a tribe in Central Mexico, they have been using many of the alien looking artifacts as necklaces. More and more of the artifacts kept showing up and soon, hundreds of images of helmets space venturing people were collected by many. Here are a few.

Not to be left out here are even more. There is no doubt that these would not be very useful helmets if one did not need to bring air to the nostrils.

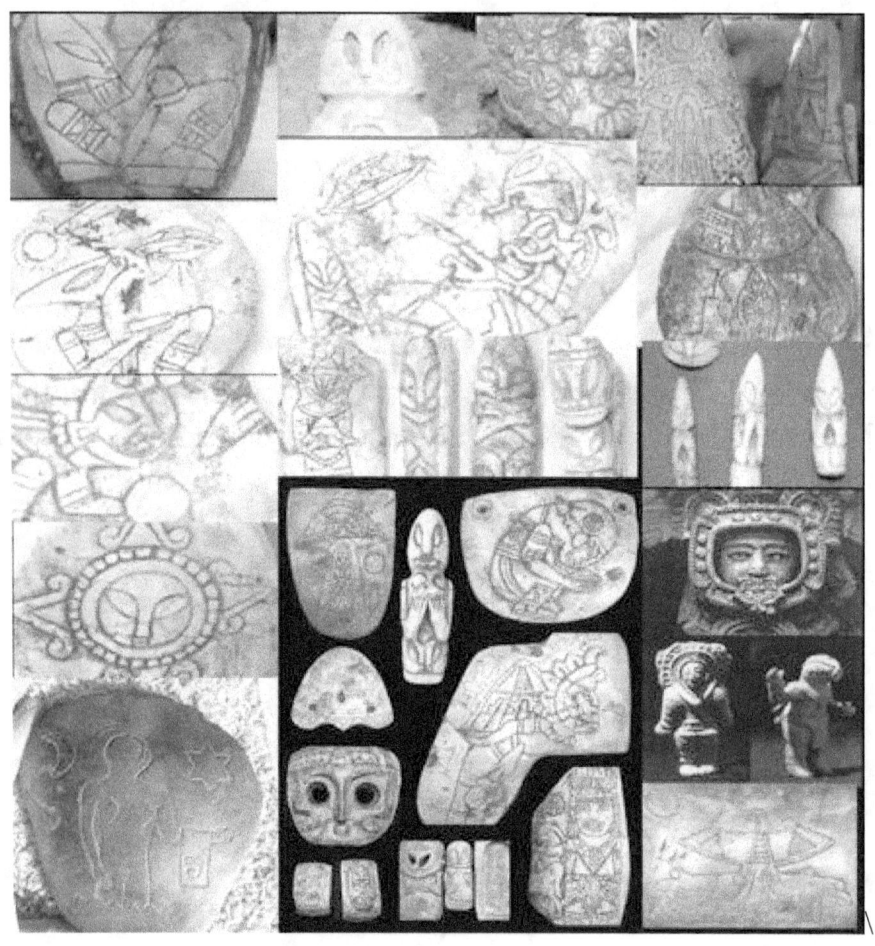

Japanese Spacesuits

The helmet designs found in the Far East are everywhere. The space suited figurines are called Dogu and there are 2 different showings; with the visor up, as shown to the right, and with the visor down. In the down position, the goggle area allows for reasonable viewing, but the level of protection is unmistakable. In fact, the entire body suit is almost always depicted as shown next.

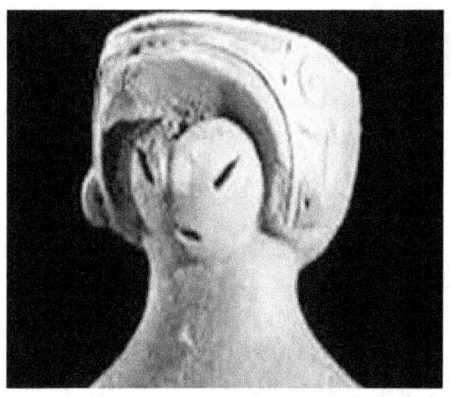

United States

This image from ancient petroglyphic art in Utah shows helmeted people with antenna coming from the helmets just like others.

Ancient Persia

The Sumerians, Babylonians, and Chaldeans all showed similar helmets and space suits. The first 3 are from Sumerian times, the 4th is Babylonian and the last is Chaldean. All depicted the same thing. One looks like a guy standing on the crescent moon and others show full body suits as if air had to be captured for some reason.

Please notice the mysterious helmet with the protrusions coming out of the neck. This looks similar to our present astronaut gear as shown at the end.

South American Space Flights

In South America we find reasonable details about space flights that should not be ignored.

Brazilian Space Flights

The traditions of the ancient Brazilian tribe, Ugha Mongulala, provide us with a fairly good representation of what we have been saying about the early time period. Here are excerpts of their traditions according to researcher Karl Brugger.

The Golden Age was a time when the gods still ruled over a vast empire on Earth.

The moon that we know began to approach the Earth and to circle around it thousands of years ago, but before the moon took its place the world bore another face. [While the idea of the moon being captured by the earth during ancient times is not probable, it is becoming more well known that the moon of Venus exploded about 12 thousand years ago and sent thousands of meteorites down to the earth which peppered most of the United States east coast to completely change many parts of the world. If you want more information, look up details on the Carolina Bays as the craters are commonly called.]

Both lands were buried under an enormous tidal wave during the first Great Catastrophe. It occurred at the end of the war between the two divine races. [This gives us a picture of devastating wars between two major groups of demigod rulers on the Earth that were warring just before the Tower of Babel Incident. This is an almost identical indication when compared to stories around the world.]

The war between the two divine races *did not only lay waste to the earth, but also to the worlds of Mars and Venus. [As we discussed, the wars escalated to other planets. This is a bizarre indication unless people knew about people living on these planets during ancient times. This ancient writing can be easily interpreted. When he said the people laid waste the planets Mars and Venus, he must have meant that there were people that flew to those places.]*

Documents were left by the gods, *which have remained hidden, they tell about the matter from which everything is made. They tell about the course of the stars and the relationships in nature. Our priests have learned how to make objects fly through space.*

Uros the Space Man

According to another ancient Brazilian legend, the Uros existed before ToTiTu, who was the father of heaven who created white men. The Brazilians were "red skinned" so I'm not sure who they believed created the Brazilians, but they indicated that this Uros character **came from another planet** which is what I wanted to bring out here.

Peruvian Tradition

In the Incan writings, the goddess Orejona came to earth in a great ship from the sky.

N. American Space Flying

We find the same thing in North America as well. According to Hopi Historian, Morning Star, in her work "Terra Papers" we learn the following about the Hopi beliefs: *The star elders were on earth when the earth was still barren. Some survivors from the war between the gods remained on Earth and replenished it.* [There is a reason why the ancient sages were called star elders. They probably came from the stars. It should be known that Morning Star got much of her details from something called channeling which means a spirit of some kind supposedly told her rather than her interpreting physical evidence or writings.]

Navaho and Pintes-The Navaho and Pintes both told of *the Golden Strangers from the skies that came in flying canoes which were armed with Burning Rays.* [Laser weapons make burning rays.]

Hopi Indians-In the "Book of Hopi" it indicates that there was a battle for the red city. *The Kachinas, which were beings reputed to be from the fourth world, came to help. In the story, some kind of tunnel was built with the speed of the wind and the Hopi were able to flee the city. The Kachinas stayed behind to defend the city and indicated to the Hopi that the* **time to travel back to their distant planet had not come.** [It is not hard to imagine that the way they left was by space ship.]

Haida Indians-According to Haida Traditions we find the following reference: *Great sages descended from the stars on discs of fire.*

Modern Flying Machines

The first group shows some of the many flying machines that have been made in modern times. While most efforts remain secret, these may give us an idea about out current capabilities to go into space even without a rocket. These are from Germany, USA, UK, Canada, Russia and even France.

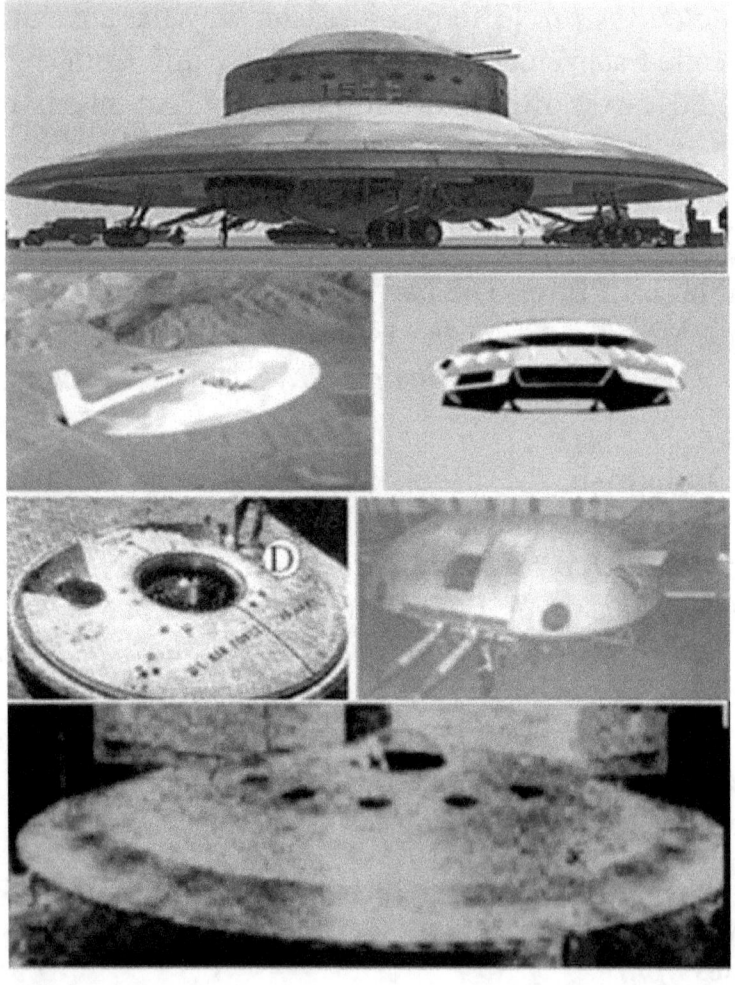

Besides those manufactured around here, we keep seeing all type of flying machines that appear to be able to go to our nearby planets Here are just a tiny fraction of those seen.

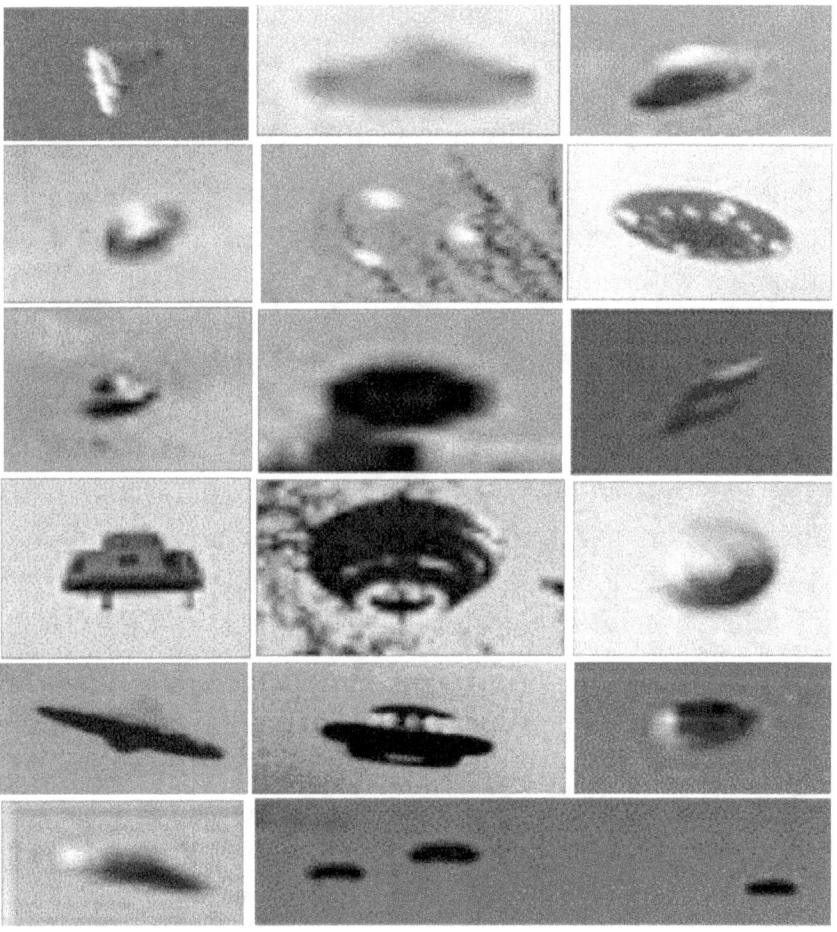

Flying machines were not always used to fight battles on Earth as we have seen in ancient texts, but are there other pieces of evidence?

Where Were the Battles?

Venusian Outposts

Probably a very active group of warriors lived on a planet that you would not have expected because you have been inundated with suggestions that Venus has always been a planet on fire. Like I mentioned before that is not the subject of this book so you will just have to trust me when I tell you that there is a lot of evidence that tells us that Venus was not only inhabitable up until it met its doom about 11 thousand years ago, but it also had colonists living on the planet. Certainly, humans could not survive on a planet that has 900-degree temperatures on the surface, but the more probable temperature was around 100 degrees or even less.

Venus may have been a target during the space wars or it may have been struck by a huge planetoid, but something bad happened. All we know to be probable is that Venus had some serious explosions on its surface, which caused some pretty nasty meteorite material to hit the earth. My book, "World War Before", described the issues that lead up to Venus turning into a fireball 11 thousand years ago.

Moon Outposts

Lunar warriors were certainly among the fighters during these ancient wars. My book, "Who Lived on the Moon", describes some of the special characteristics of this special place.

Other Places Besides Mars

Besides Mars, other planets and moons in our Solar System seem to have at least been touched by humans. My book "Evolution of the Planets" covers some of those interesting characteristics. That brings us to Mars.

Martian Outposts

In the following sections I will show that life was on Mars in the not too distant past and the inhabitants of that planet could have been involved in the Earth wars. Not only can we find the evidence that mankind lived there, but also that the people fought on the planet at some time in the past. Possibly there are still some alive and visiting us from time to time. Do the following images of UFO pilots look Martian to you?

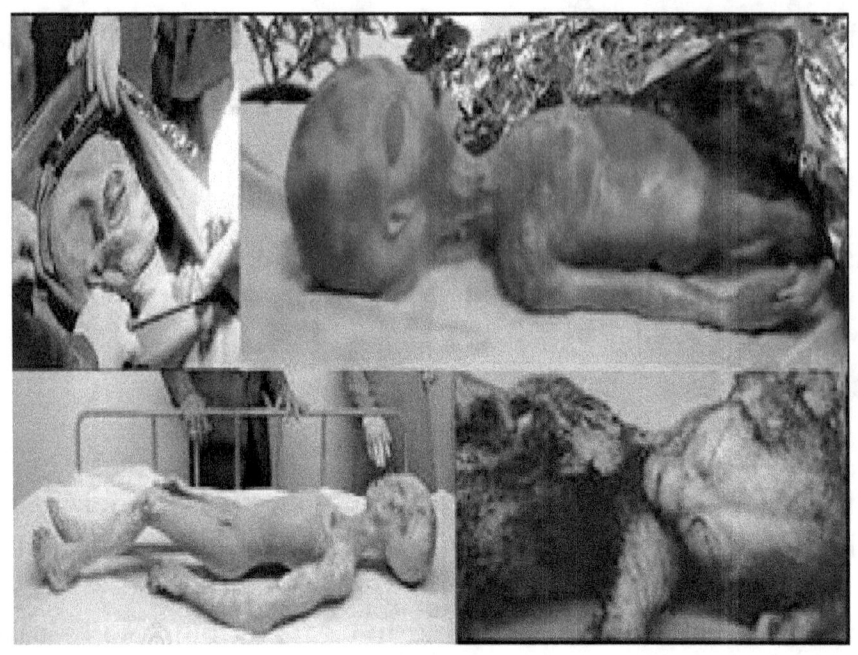

If those recovered UFO Pilots don't look right, how about this next set.

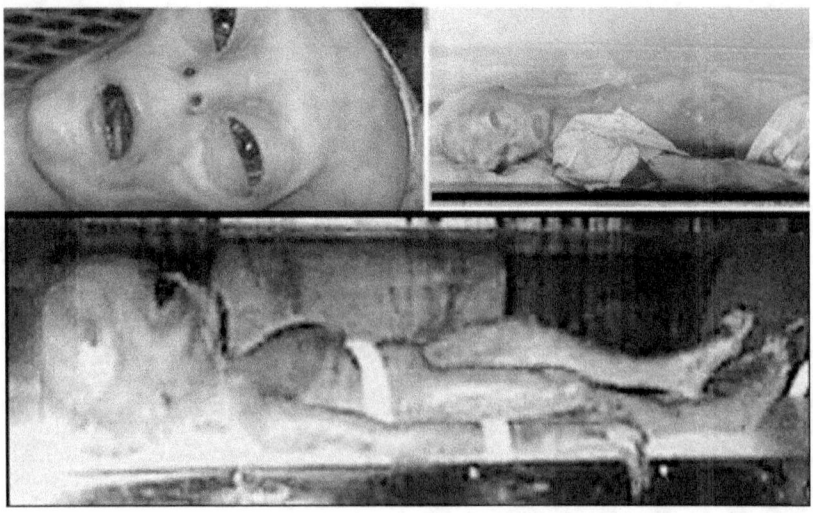

To see if someone could have lived or even lives on the planet Mars, first we need to look at atmosphere.

Martian Atmosphere

It should be noted here that there are planetoids in our solar system besides earth that contain oxygen in the atmosphere. Those with the most and best were Mars, Venus, and Europa, one of the moons of Jupiter. Sometime in the past; Mars began to lose its "normal oxygen" atmosphere and its surface available water. Finally, all the inhabitants left or were destroyed. We know that Mars had a substantial atmosphere in the recent past because the surface of Mars has evidence of substantially fewer craters than found on the moon and the ruins of many buildings are still visible today. That doesn't mean that there are no craters of Mars, in fact, there are about 2700 that are over 20 miles in diameter and 15 that are more than 130 miles in diameter, but **almost all of them are on one hemisphere of the planet** [the southern hemisphere]. The percentage of craters to landmass on Mars and Venus are both substantially lower than that found on the moon because of one thing. Recently they both had substantial atmospheres. Venus still has its atmosphere, but the Martian atmosphere is almost completely gone and the stuff left is almost all Carbon dioxide. Who wants to breath that stuff?? Here is the probable truth. Before some terrible changes occurred, Mars was inhabitable by humans and there is evidence still visible that this inhabitable place was inhabited.

Timing events on Mars is extremely difficult but the idea that humans lived there removes some of the anomalousness of the evidence that can be found. The most likely reason areas look like cities is that they once were cities. The reason that these areas looked like they were burned out, most likely is that they were, and the reason the buildings look similar to earth buildings is that earthlings most likely built them. Humans molded Mars and, most likely greed or someone trying to increase his "power" caught Mars unaware and soon it was too late. The atmosphere, the rivers, the lush vegetation, the laughing of children almost completely disappeared. When I say almost, there are surprises still to be seen.

Martian Cities

There are many signs of life on Mars and many anomalies that strongly suggest that this was human life. Possibly, a majority of the population was destroyed during earlier fighting in ancient wars, but surely some remnant survived. This goes along with both scientific reason and religious record.

From photographic evidence and from research initiated by researcher Richard Hoagland, the remains of human existence can be found everywhere on Mars. Much evidence is covered in years of dust, but is still recognizable. The picture following and associated drawing following shows the remains of only one of the seven older "cities" that have been "found" on Mars by the various picture taking probes that have been sent in the recent past.

The drawing tries to show what the city might look like with the major streets uncovered. Even with a substantial amount of damage to the cities, there still is a significant amount of detail including right angle walls, pyramid structures, and walled courtyards Note the almost perfect, huge pyramid [second building from the right].

Let me say that my pictures do not do justice to the actual finding of the various researchers on this topic. My desire here is to let you know about the findings and provide an overview. Not only does the above "city-site" seem manmade, but other areas have possibilities of showing signs of life as we send more probes to our closest neighbor. More details concerning Martian habitation can be readily found through details collected by many sources. Like I mentioned above, the premier investigator is a man named Richard Hoagland but there are hundreds of people who have looked over the photographed areas of Mars and other planets and have reported on what they saw. Don't just look at this brief overview and think that this is the only evidence of human existence on Mars.

I am only presenting a small quantity of elements just to show that there is enough evidence to even allow the most ardent "non-planetary habitationalist" cause to wonder. The photograph below shows the possible city site and the now familiar area known as the face on Mars. Whether or not the apparent face is a "manmade object or not is not known by the author, but it is interesting to note how close it is to town.

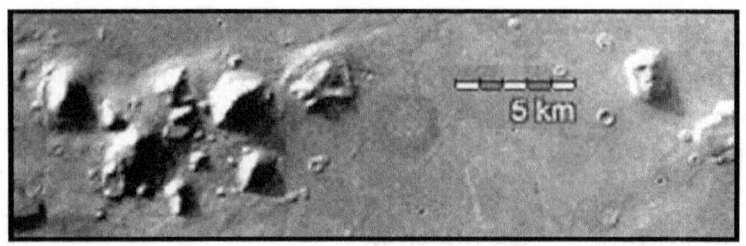

Speaking of peculiar, what researchers have found is that this "city" is right on the border between to smooth side of Mars and the cratered side of Mars as shown in the map following.

Ancient City Number Two

It is improbable that a city would have survived the moving of the mantle on Mars after the near collision with earth. Therefore, the most likely scenario is that this dwelling area and the others nearby were all built and inhabited well after the 400-thousand-year-old date. We can believe some colonists were here as late as 6 thousand years ago.

The second "city" shown below contains many well defined pyramidal "buildings" that are positioned in a matrix that resembles a city. The pyramid in the upper right is especially interesting in that, even the steps up to the building can be made out. The drawing below may bring out some of the features. By the way there are better pictures of this area and the other areas shown. They can be seen in books by other researchers. My mission here is to simply bring up the probability. I placed lines on the picture to signify major roadways.

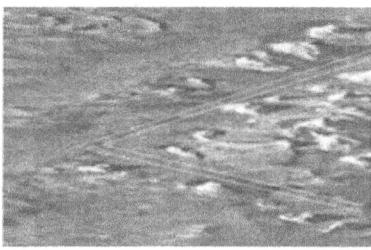 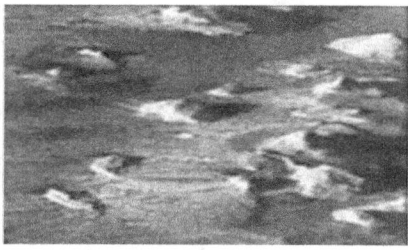

More Towns and Cities

The next city seems to have a perfect stepped pyramid at the upper right. I have drawn its general shape to the right for clarity. Several steps and platforms are clearly visible.

Still another set of remains looks like a city once was along the seashore.

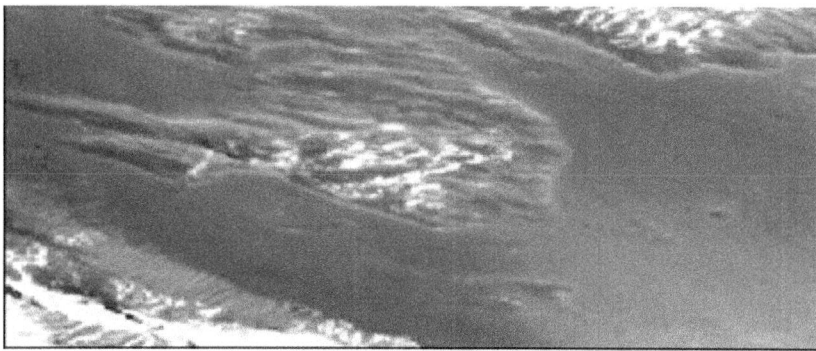

The last image in this section seems like some great amphitheater with buildings around the one end, possibly as an ancient portion of a town.

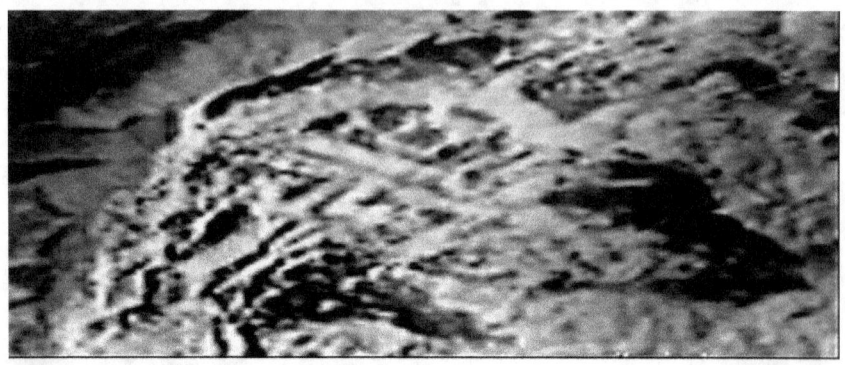

Buildings Galore-Speaking of right angles; look at this pentagon that was found on Mars. Several steps and platforms are clearly visible.

Pentagon-Like our own pentagon, this building certainly describes the greatness that once was had on the planet Mars.

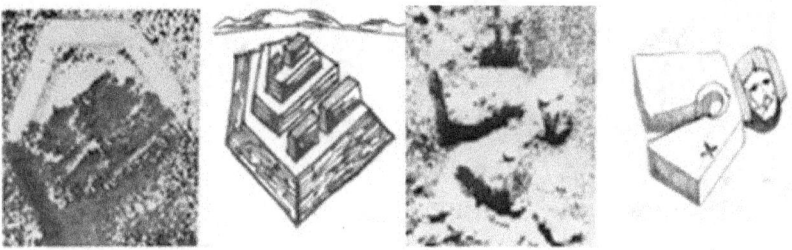

Library-I don't have any idea if this was a library or not, but it seemed to has that austerity of a place with books so that is what I call it. [See above right] Can there be any doubt that these some of these structures are man-made? That Pentagon thing was first discovered by Richard Hoagland and it still impresses all who see it.

More Pyramids-Pyramidic buildings have been found everywhere on the Martian surface. Some of these pyramids are 4 sided like those seen in Egypt while others are three sided. As shown in the upper right corner of the following picture, one area on Mars is covered with many, many triangular pyramids.

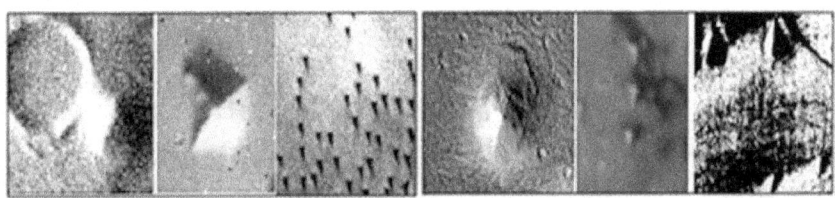

More Square Buildings

You don't have to look far to see more square building either. Look at the blow up near one of the Martian pyramids. Notice the building is completely square. Try to make this pattern with wind and erosion. Notice also from the shadow that the building is fairly tall and the top is smooth.

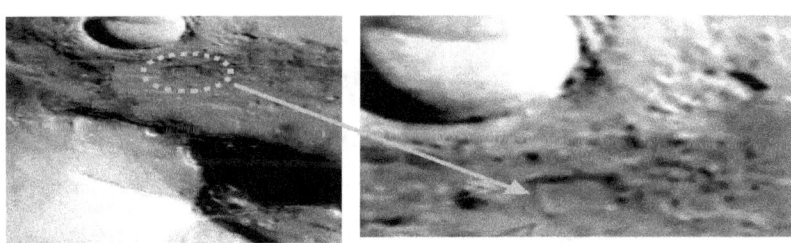

The Famous Martian Airport-If those photographs are not enough to convince you; look the picture from Mariner 9 that Richard Hoagland termed as the Martian airport-out by itself, as you would expect, but with unbelievable detail. The, Tom Penner, rendering of the complex certainly shows that, at one time, this was an airport or huge shopping center or something else, but it was not an accidental structure. Actually, the structure seems to be mostly underground rather than above ground as the artist conception shows, but you can get the idea just the same.

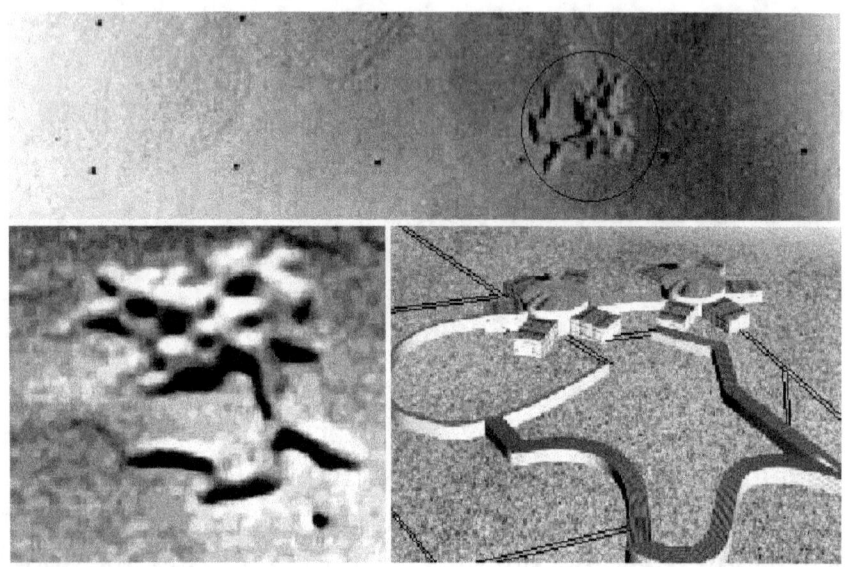

Huge Courtyards- Speaking of symmetry, look at the right angles, long straight lines and height of this area.

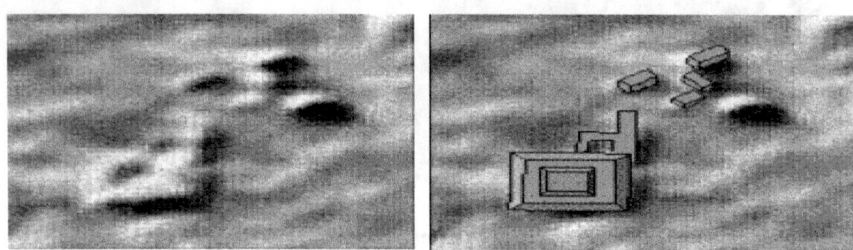

Perfect Hexagonal Building-Speaking of manmade, the Hexagon pattern is not usually manufactured by wind, rain or dust. It is usually manufactured by a human. The following shows such a building on the Martian surface. This one is interesting, in that, the inside and outside wall can be seen showing regular thickness of the structure around the entire perimeter. It almost looks like a short hollow hexagonal tower.

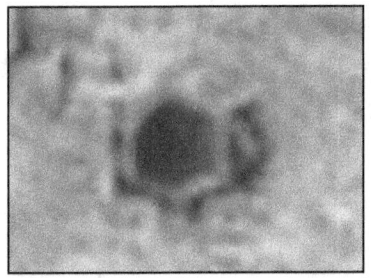 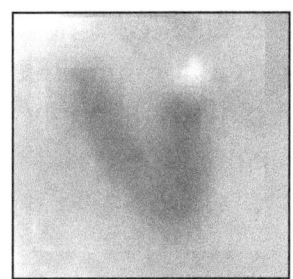

Towers-Some of the towers on Mars are just tall and have no fancy geometric shape. The one shown is just a run of the mill cylindrical type. Slightly wider at its base, it rises up hundreds of feet into the Martian sky. In the next image we can see a major tower still standing. This one seems to have a base complex below with long corridors to the huge tower. Protected from the environment by the nearby rock face, it has stood the ravages of time fairly nicely.

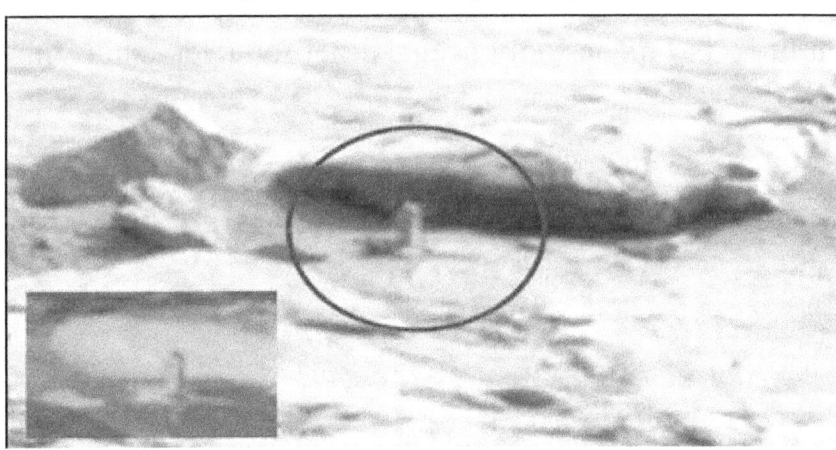

Other items have also been found; each with its own level of curiosity. The following collage includes several of the more famous ones. From the left and across are a dog house, "Giant T", six sided perfectly formed pyramidal structure, a structure that looks like a rabbit, a structure that appears to be an arched gate, a strange building with a rounded end, and fine detailed lines in front of a structure that looks similar to a dolphin. These, by themselves, might not be the evidence of civilization; that is true, but the huge amount of detail begins to sway even the most cautious observer.

Still more items showing the remains of some civilization keep turning up.

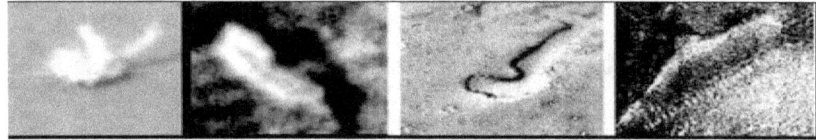

Besides the massive complete cities that are evident, there have been found enormous amounts of artifacts that tell us that life on Mars has only been extinct for a very short time.

In this section showing an industrial community, even when a section is blown up the walled building, streets, towers, and rectangular buildings cannot be ignored as a human inhabitance.

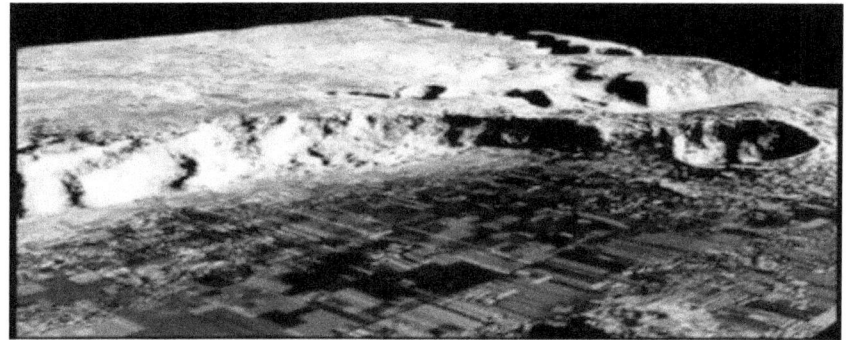

The above NASA image has been enhanced below to show what appears to be buildings, streets, and all the rest.

If that was all, we might be wondering, but here is another and another on Google Mars.

At another location we see this! Hidden next to massive cliffs we find buildings streets that look like a city. Next to it is another ancient industrial center.

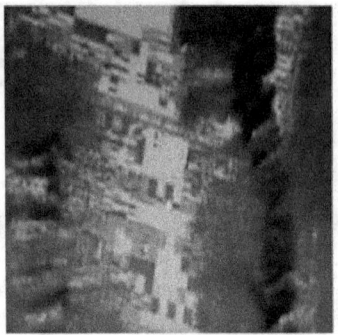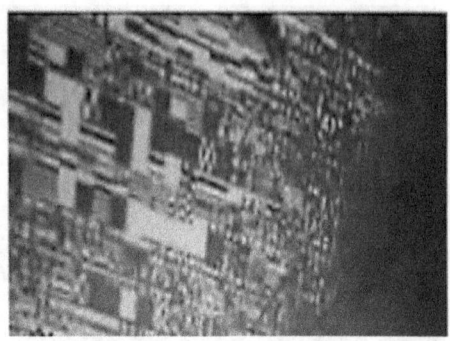

At still another location we find another similar section. Even the windows on the buildings appear to still be intact.

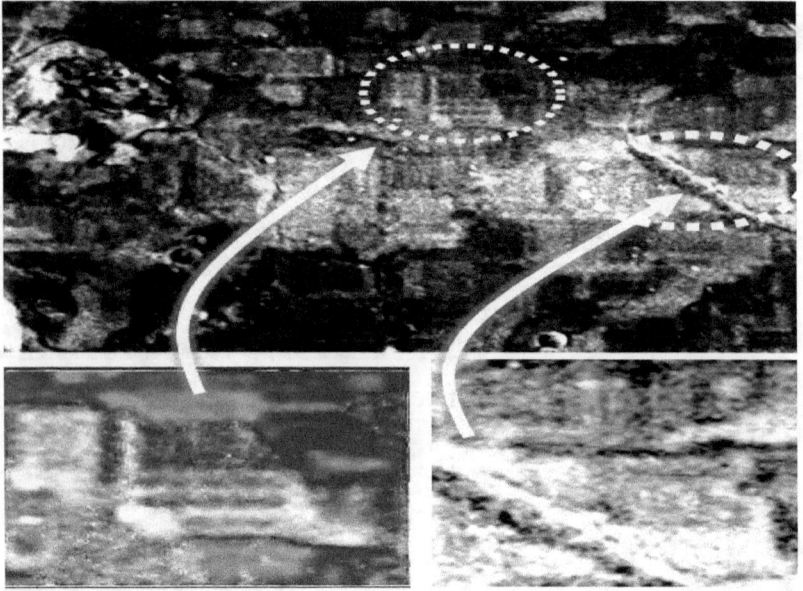

What you can see from the image is that the area is almost identical to a rural scene on the Earth to the right except for the occasional meteor crater. Below is a rural area from our planet. The streets are more defined, but the particular sectorized image seems very close to that shown for Mars. Except for the gash down the middle of the town, this could be some earthly business area.

Still Another Industrialized Section-This comes from an unassuming area known as Hale Crater. The tiny square on the picture below shows the section of the protected crater that will be blown up for viewing. If we blow up this section, we find that the above details are certainly not the only indication that Mars once was a hugely industrialized planet. Details are shown next. Does this look like an industrial section or a dead planet?

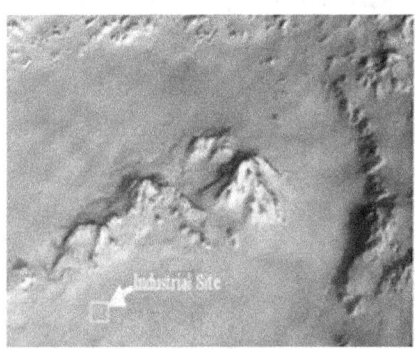

If we blow up the section on the left, we can see all sorts of buildings with rectangular gridwork and parallel walls, and regularly spaced building separations. There can be little doubt of the massive structures in this area. [Above right]

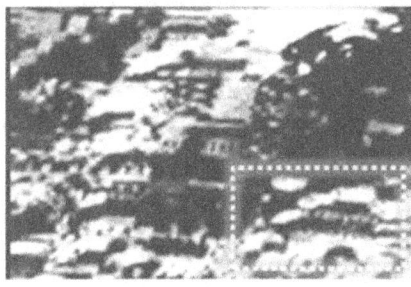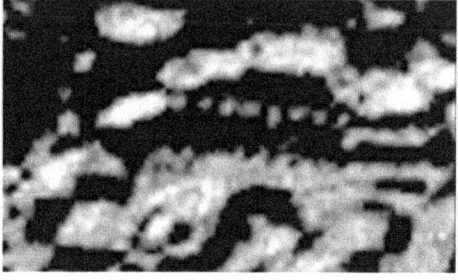

If we blow it up more we can view one of the most magnificent cultural centers on the planet. Notice the eight visible columns in the front and the multilevel terraces, the various pools [now empty] and round silo looking water pressure tank or similar functioning tank. These all appear to be fairly modern which gives us one level of understanding concerning colonization in the no too distant past.

We could go on and on looking for more signs of human occupation in this section, but I think I will let you do that on your own as we look other places. I have no idea what this next thing is, but it is not naturally formed. The massive 4 balls in a row thing and the arm coming out to touch the thing along with the bridge like artifact that all this stuff is on could not be naturally occurring. The image to the right is some type of tower, but this time some massive light source is shining out over the Martian surface.

The following set of six square things was probably made by nature, but that would have been some Martian's name. Good old nature loved to make weird things. Possibly, this is the remains of some ancient port, but not much is left. To the right is a square thing almost completely covered by the sands of time. While it is thousands of years old, there is no question a huge square structure once was on the Martian surface at this location.

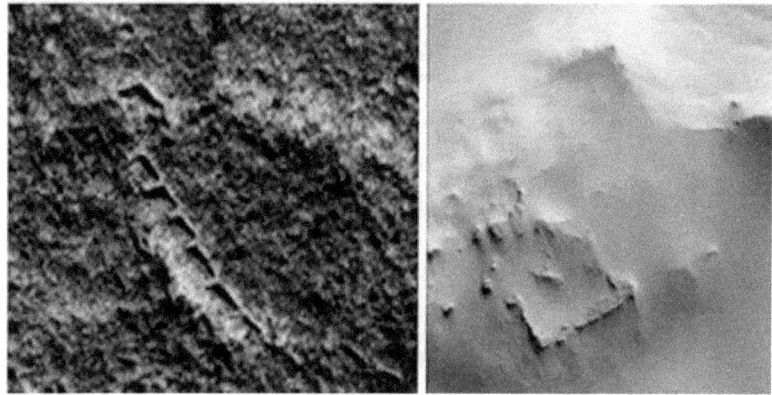

Triangular Somethings-I don't know what it is, but I do know that it is triangular, symmetrical, has very straight proportions and is huge.

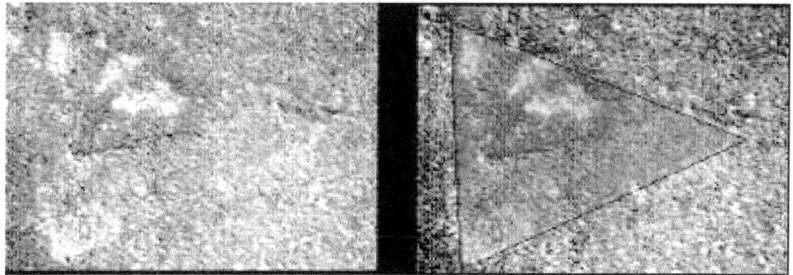

Martian Sphinx-From the Martian Lander, one of the pyramid structures could be seen. The lander also viewed, what is believed by some, to be something similar to the sphinx. Either they were copycats, or we were. If we think of time frames and the theory that the Great pyramid may have been made about 30 thousand years ago, this may be an indicator of the time when these structures were put in place.

The elements might not initially be seen, but when you add a few lines and definitions, the sphinx and pyramid become clearer.

Strip Mining? The picture following left, is not where someone drew a spiral in the ground. This thing is immense and looks very similar to modern strip-mining operations where the sides of the hole are used as the ramp to bring materials up from the bottom. If this is not a strip mine, what do you suppose it is? I can assure you it is not naturally created.

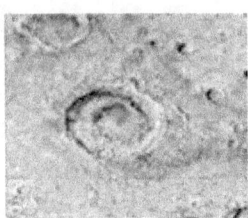 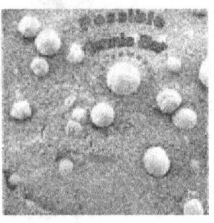 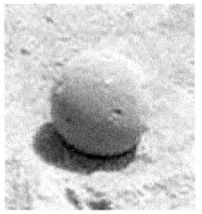

Tennis Anyone? This latest finding by our Mars rover has me wondering about sports. Perfectly spherical balls seem strange. I believe that the one in the middle is a tennis ball, don't you? No matter what they are, these things do not look like something that is made naturally. Wait just a minute. Did we find another perfectly spherical ball at another location? [See image right above]

Valve-The next image in this series is of a valve. I don't mean something that just looks similar to a valve, I'm talking about a cutoff valve buried just up to the surface. Do not even imagine this to be some trick of the eyes.

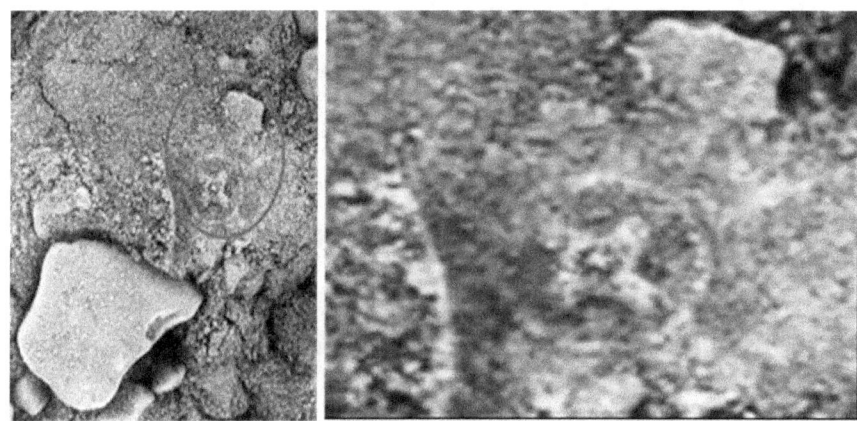

Even the bolt holding the handle on can be clearly seen in this blow up. Maybe other aliens in the universe make the same type of valves as we do, but it is not likely. Some might think the people are from here.

Funny Things

The next group is just for fun and we see what appears to be a tennis shoe and a Sasquatch. I have no idea what the last thing is, but it is very strange for a rock. Possibly it is an old tree trunk.

Yes! You are right. A Pith helmet is lying on the Martian ground in the following image.

The next collage appears to have discarded, man-made items: including the end of a pitchfork, the leg of a couch, a huge spinout that is 80km long and does a 360^0 spin, that is followed by a simple piece of metal and a perfectly bored round hole and what looks like a giant image of a Spade from a card deck.

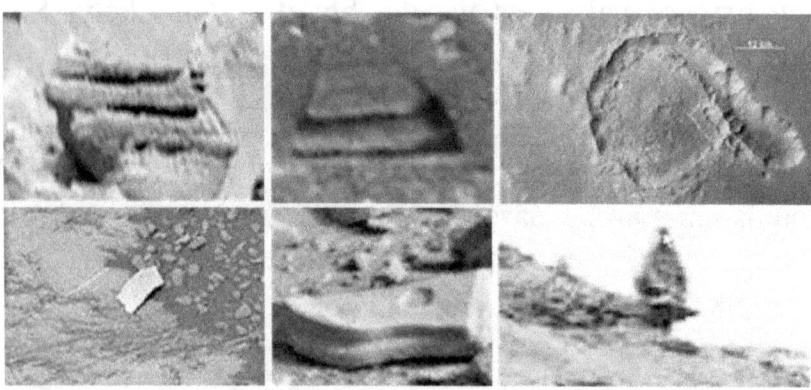

After a giant spade image, we look at puddles and worms.

Puddles and Worms

Puddles of the Martian surface look more like the surface is very hot than very cold. Whatever made the surface puddle is a mystery and here is a picture to ponder. To the right of the puddle picture following is another Martian anomaly. This looks like long rows of worms, but actually, the se regularly shaped hill features don't seem to fit normal land masses. Their rolling shapes appear to be dunes. [The 'worms comment' was to keep you interested.]

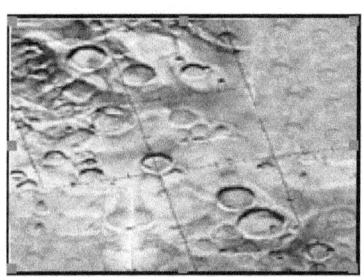 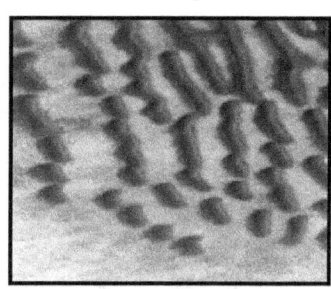

Martian North Pole Things

This is really something to see. When we look north, we find extremely unusual billboards or graves or markers of some sort. It is difficult to believe that these tall structures, shown below, could have been formed naturally. Whatever these huge billboard objects are, they are not covered over from thousands of years of existence and must have been made fairly recently. We could estimate that they have been there no longer than a million years ago and probably much less time.

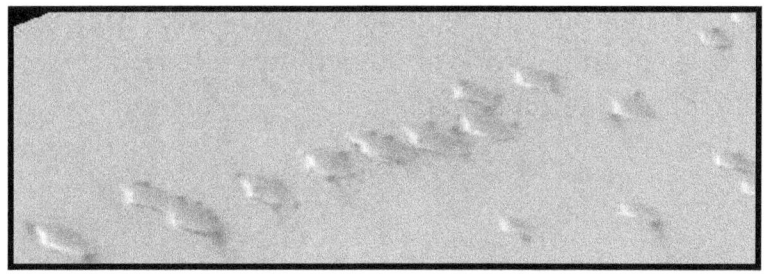

Still More North Pole Things

One way the Martian surface is being changed is by what appear to be tilted trees. [left below] Actually, they look more like the huge effigies found on Easter Island. All placed at the same angle. Like the strange billboard structures shown earlier, these tilted things are all over the place.

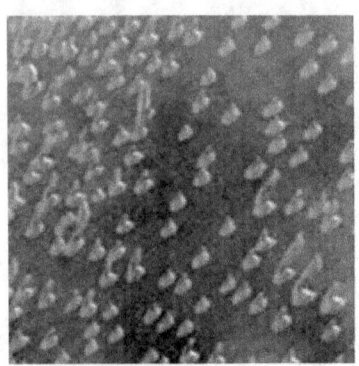 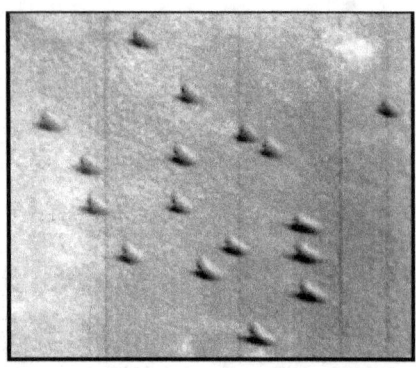

Another way the Martian surface is being changed appears to be with Hershey Kisses [right above] as this picture from its North Pole shows. Like the slanted things, I don't know what they really are, but we don't have them on earth, so I thought you would like to know that they are on Mars.

Q-TIP

The next strangeness I call the cotton swab. This massive suspended arm could not have been made accidentally by Mother Nature, but I have no idea what it was. Maybe it was some type of tuning fork or something is buried underneath that would give us an explanation.

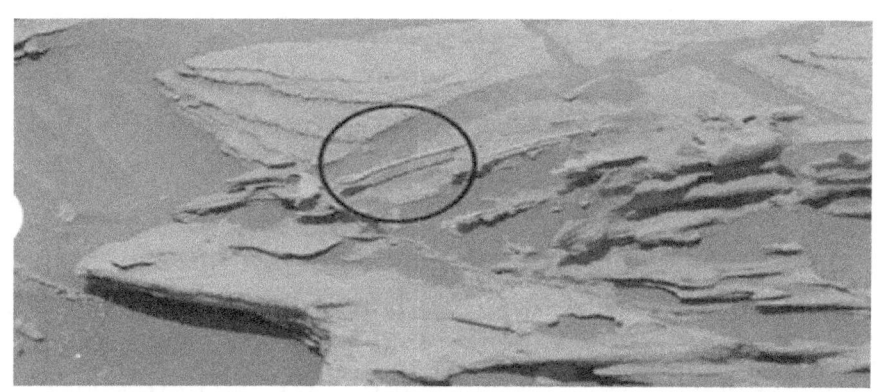

Flying On Mars

Certainly, there might be ancient satellites on the Martian surface from ancient times, but there can be little doubt that flying "things" are in the atmosphere of Mars. Here are just a few. Notice the shadow on the ground of the third one.

The next image seems to show some type of stabilizing Rutter.

Here is another just for good measure. Like the others whatever is flying around is seen often.

Crashes

Our little robot on Mars may have seen a piece of some type of machine lying in the desert. To me, it looks like the tail fin of some flying craft. Maybe, it was one of the things flying around on Mars. Please do not believe this is naturally formed.

The next image appears to be similar to flying saucers seen on earth.

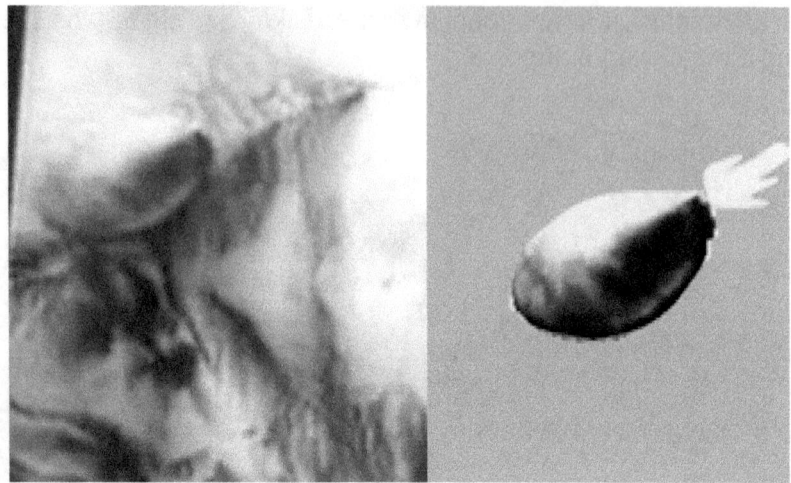

As we look farther, another crashed airplane or something very similar can be found. As shown next.

The next image left even looks like one of these things is flying over the surface of Mars.

Beacons

Even if someone lived underground, occasionally, one would think we could get a glimpse of the underground lighting. The next couple of images seem to show just that.

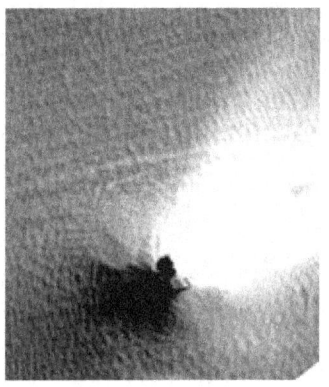 

Martian Roads?

Lines Mean Something-What are the lines in the Martian picture following? The lines go miles and miles, traveling over the top of crater blast areas. The set of furrows on the Martian landscape pictured extend in an absolutely straight line. No naturally occurring phenomenon can account for the lines that continue over mounds and through troughs and we know that they were made in recent times because the meteor blast marks are below the furrows and none are on top of the strange lines. It almost looks like a huge farm. The deepest furrows are towards the center of the picture, but you can see many more furrows below the major ones all extending in the same direction. No one knows what they are, but they were made very recently.

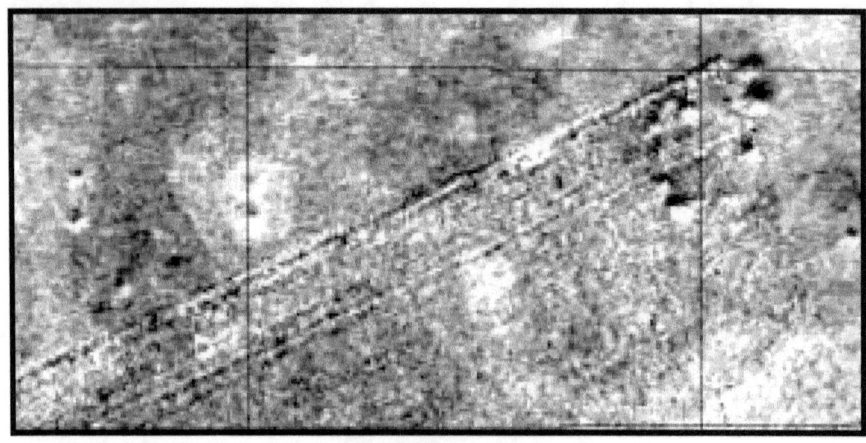

Long Roads

The remains of roadways can be found all over the planet. Here are a few of the indications found to date. Note the extremely long distances and almost straight avenues.

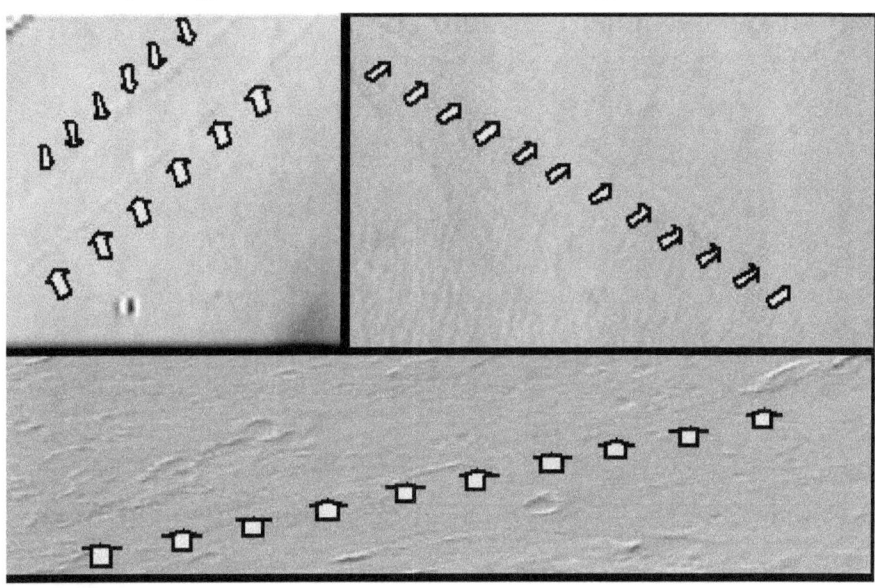

Note also parallel avenues discounting most geologic possibilities. Note the finely defines lines which would correspond to roadways.

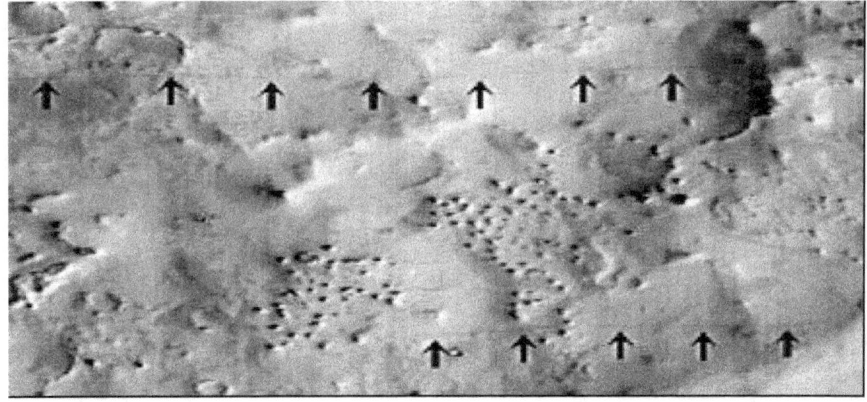

The following image shows some type of vehicle going across the Martian surface. If you look closely, you can see there are actually 3 or 4 sets of tracks so this is a well-used highway. OK! I mean well used for surface travel on Mars. Please notice how this rover and its track and the one I pointed out on Lunar look similar.

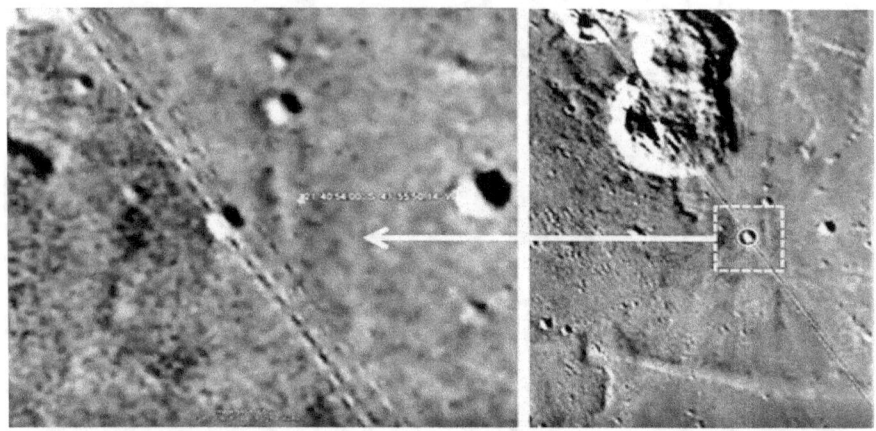

On a similar roadway, we find something decided to go a different course.

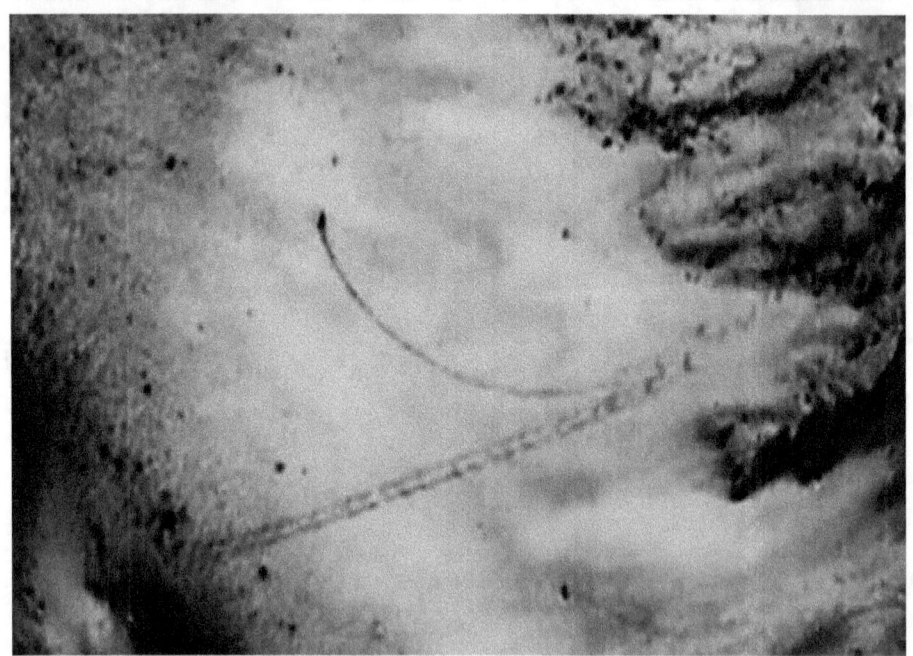

Could There be Vehicles on These Roads?

Certainly, light can play tricks on what you see from millions of miles away, but the next two images seems to look like vehicles of some kind. The second one even shows its "track" or what appears to be one. Don't ask me what is on top of the vehicle either.

 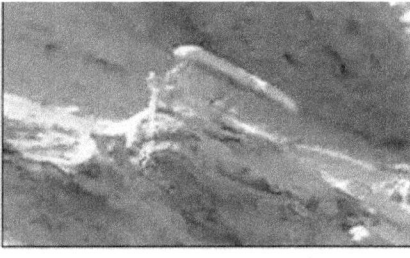

Maybe the vehicles are out looking for something very precious on Mars----water.

Martian Water

Something happened on Mars that destroyed everything and soon the atmosphere left. That event was not the near collision with earth that left half of the planet blown away and the remainder pitted with craters. That earlier event, most likely occurred about 400 thousand years ago. Even after that catastrophe water was still abundant on Mars and with water there was, most likely a reasonable atmosphere. At a later date, the atmosphere began to thin and the water began to leave.

Although we don't know the exact time period for this final disaster for any colonists, the picture following shows the remains of a huge sea on Mars. The remains are not very old and the details are still distinct, so we can be fairly certain that it was not very long ago. Some researchers place the rippling of the ground at about 17 thousand years ago by its strong definition. This may make sense. Later we will look at an ancient document indicating the Destruction of Venus and Mars happened at the same time. Venus got the worst of it, but possibly 11 thousand years ago, Mars became a ghost-town----I mean Ghost Planet.

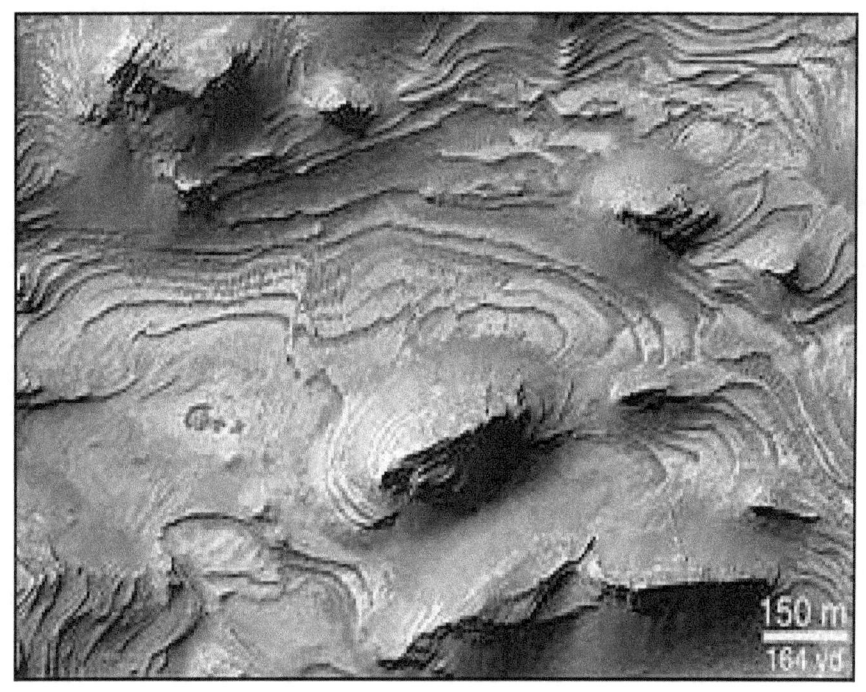

Liquid Water

Speaking of the rippling of the ground and a 17 thousand year ago loss of water, here is another twist. From studies of Martian meteoric material, Dr. Leshin of the University of Arizona and others have concluded that Martian water originally contained higher deuterium levels than previously thought and the Martian atmosphere has lost two to three times less water through the eons than "dry planet" models suggest. Therefore, some have concluded, there must be a huge ocean-like reservoir of water beneath the planet's surface. Not 17,000 years ago, I'm talking about water on Mars today. The picture below may be one of the damp areas on the surface today.

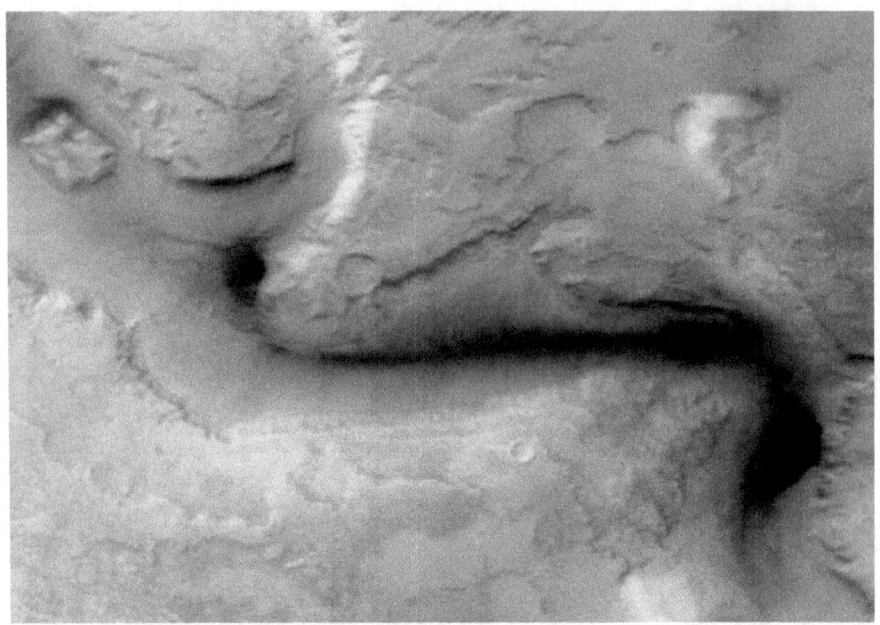

The recent picture below shows a reflection off of a shiny surface. It also appears to be a frozen lake on Mars. To the right you can see ripples in the water captured in a crater.

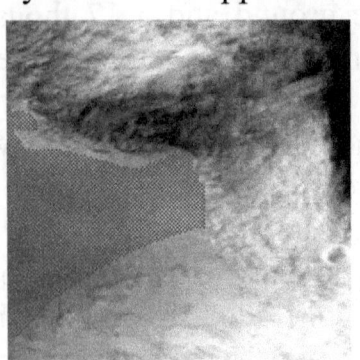
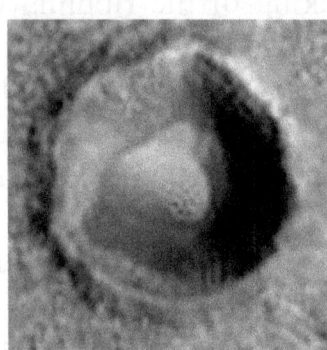

Rivers

A good example of a dry river on Mars is shown below. There can be little doubt that the water flowed along this path in the not too distant past.

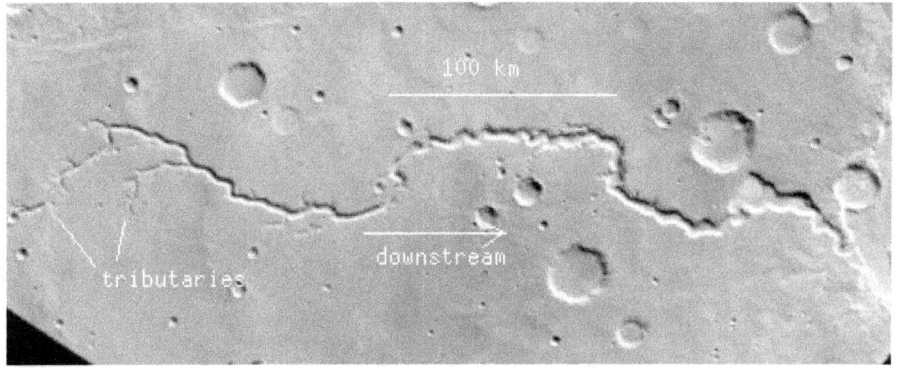

The next is the remains of a massive river delta no longer flowing with any more than the tiniest bit of water, but, at least, it's something.

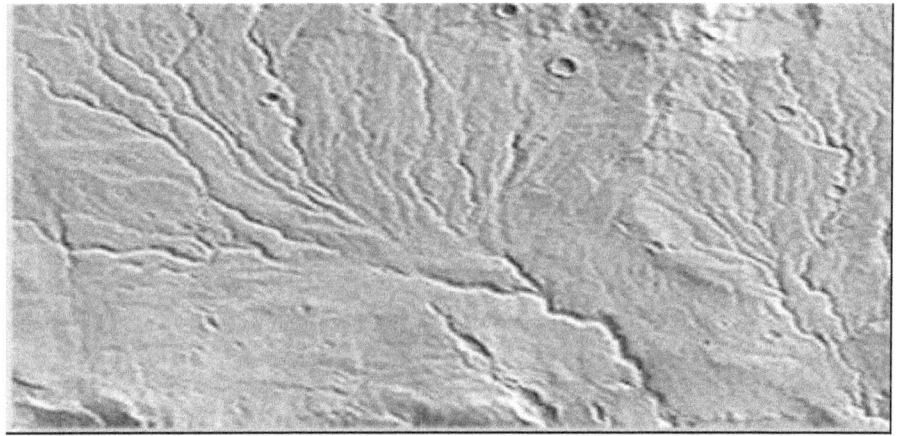

Sand Dunes

Look at the following picture. It appears that on the northern part of the planet, there must have been a substantial sea which produced thousands of sand dunes as it slowly decreased its mass. I don't know the details of how it all left but there surely was a massive loss of water leaving only the sand piled up from the wave action. If these are dunes, the water would have been on Mars in the fairly recent past, or the structures would have been reduced to hills.

Martian Sea Port

The picture below shows what has been reported to be a Martian Seaport. This area is commonly known as the port or seaside retreat. Note the extremely regular shapes associated with the building on the cliff that once could have been adjacent to a waterway below. I have superimposed a drawing of the building for clarity. The right-angled surfaces of the building are still very distinct showing that it was not abandoned extremely long ago. I know the three squares with round areas on the center of each looks like some odd-looking home, but think of the guy that owned this retreat as an eccentric.

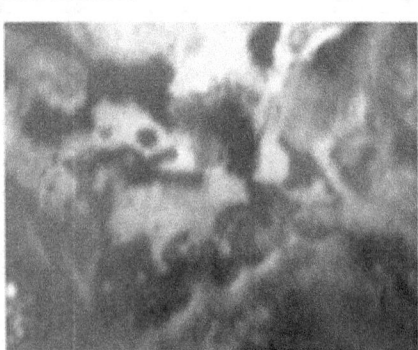
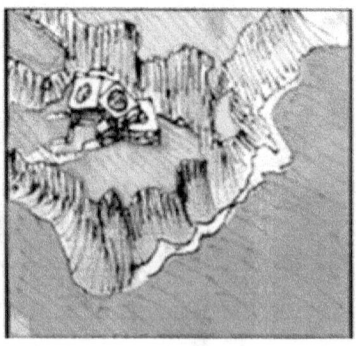

On the following image shows the area where the Martian Seaport was found.

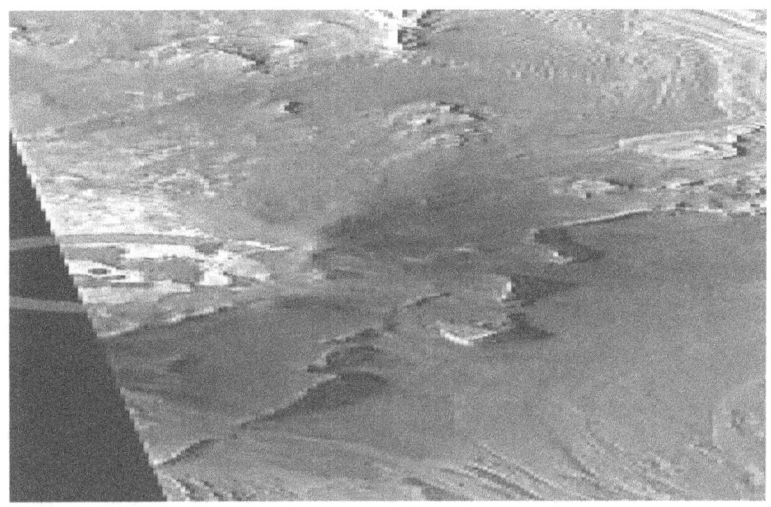

Lakes

The previous abundance of water is evident everywhere on the surface. Here are 2 areas that used to be lakes, long emptied of their water as Mars lost most of its atmosphere, but one can still see what appears to be some vegetation so that one can image some water still remains in this area.

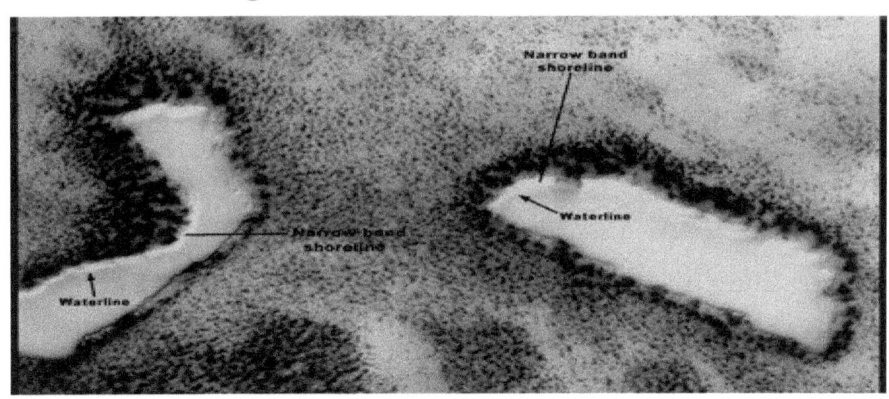

Something Interesting

I'm not sure what the next object is or was, but it certainly had a massive dome at it center. Well away from any city or other element, it is just another anomaly that appears to be man-made. The first image below shows the area and the second one I have blown up so you can see the detail.

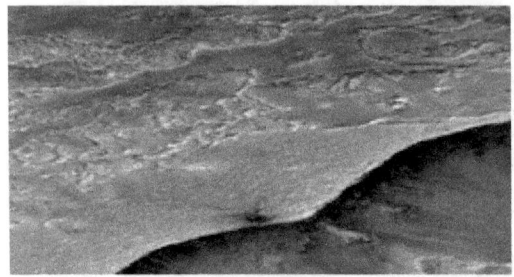 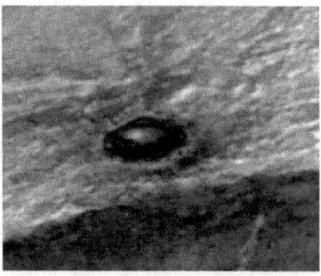

It looks like the next few images actually are flowing a little.

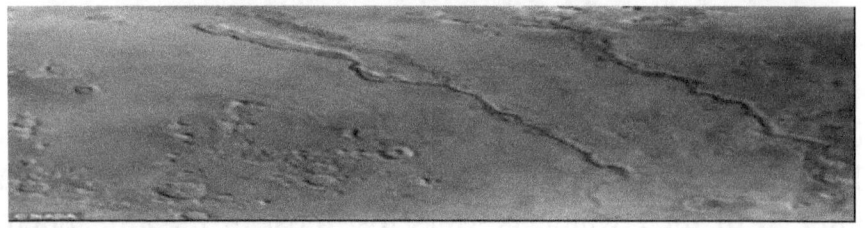

Without a doubt there are signs of water all over the place. Here are a few more.

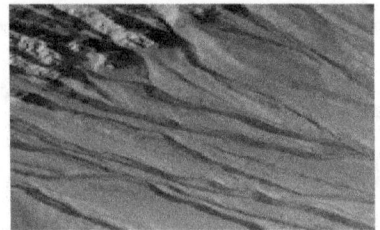 

Where there is water and an abundance of CO2, you will find trees. With 95% of the atmosphere as CO2, we just have to find trees!

Martian City Destruction

Not only do we find cities, terraforming possibilities, and all types of manmade objects, we find evidence of war or something very similar to it. The cities of Cydonia and the others that were presented earlier have all been deserted, but the specific details showing massive wars on Mars are not well defined in those area. However, some of the municipal sections found on Mars are not hiding probable massive wars. We see some cities with complete destruction. Look at some more of the curious details found by the many researchers in this area. Can there be any reasonable doubt that these structures are regularly spaced, rectangular, and have high walls just like you would expect in a city. Even what appear to be sectored roadways can be made out. The most striking element is that the buildings appear to be melted. The buildings are not simply covered up from the ravages of time; they were burned up by some type of weaponry like nuclear bombs. Look at the remains of the building in the upper center. Melted, melted, melted and the people that were there probably looked like the ones found in Mohen jo Daro on Earth.

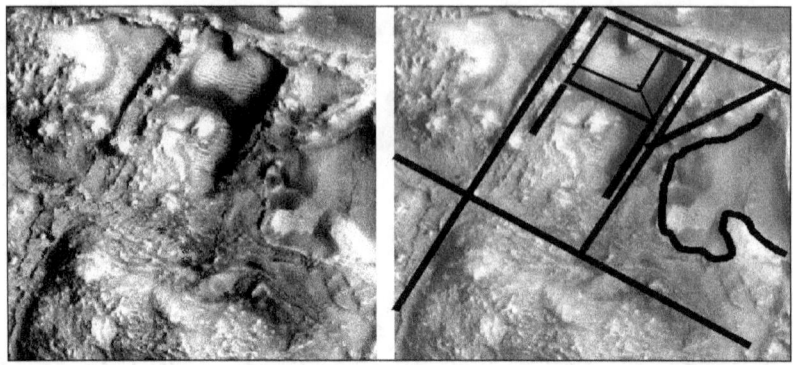

In this next picture, you can see the crumbled walls of the city pretty clearly. It looks almost like the city had been in a war. The roadways are pretty clear, but the walls of the buildings are all blasted. Parts of the walls are still standing, but clearly, something horrible happened to this city besides being buried in the sands of time.

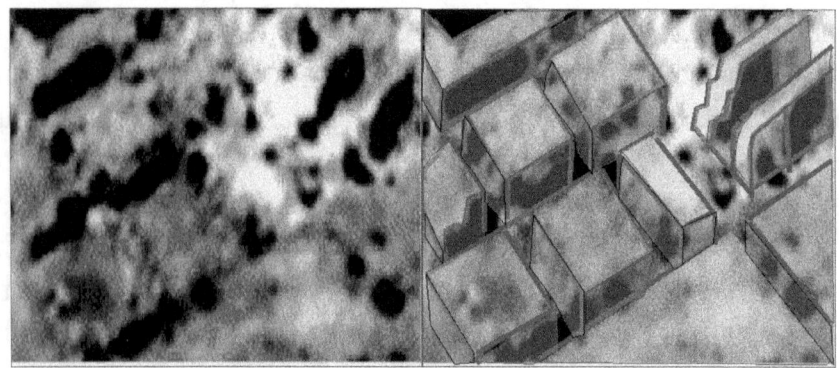

Another Destroyed Complex

Near the Huge gash in the planet, Valles Marineris, we can find still another destroyed city, village, or complex. The larger pyramidic structure seems to be partially intact, but much of the rest is almost completely gone.

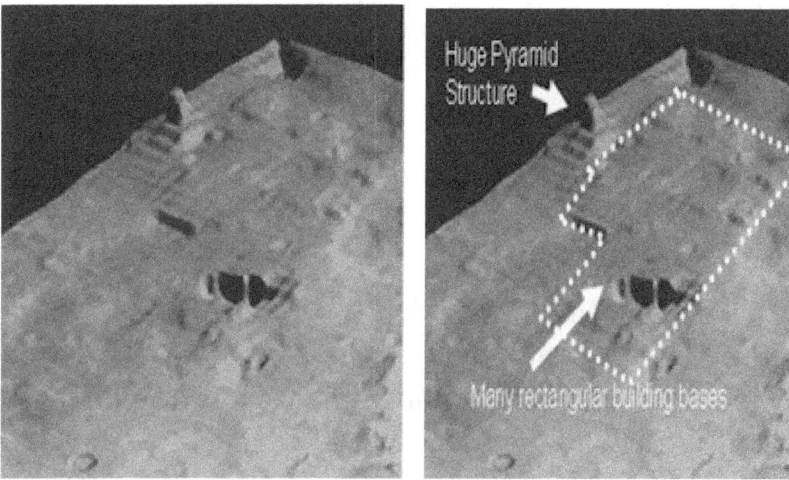

Still another burned out remains of what possibly was a great fortress at one time is shown below.

As we look at these ruins, there are two things that seem to be evident on the Martian surface. Some high-powered bolts of lightning-like terrors hit, nuclear bombings filled the Martian skies-----or both.

Lightning Nukes

Today we are finding evidence of terrors we can only imaging. There is not many who contest the evidence of some type of nuclear bombardment and something like Lightning struck all over the surface of Mars.

Xenon-129 Evidence

One way to test for nuclear war is by looking for radioactive remains such as Xenon-129. This Xenon-129 stuff is a "second order nuclear fission by-product" and guess where too much is found. Mars has nuclear by-product in abundance. The problem is that no one can determine how nuclear blasts went off on Mars. Don't believe stories of Mars collecting the material from some ancient supernova because it just doesn't make sense. What does make sense, is nuclear bombs, but then there would have to be people there and someone must have been mad at those people. Many of the scientists trying to figure out why this Xenon 129 is there don't believe in the very recent nuclear explosions on Mars, but they are completely baffled as to what else might have caused the quantities of Xenon-129. That type of stubbornness keeps our history books comfortable and useless.

Lightning From Nowhere

While there is little atmosphere on Mars today, scientists found huge lightning rilles on some of the planets including

Mars. Early writings indicate that lightning weapons were used during the horrible wars. Maybe these lightning weapons were used on Mars. The unusual zigzagging ditches can be found on earth after huge lightning bolts strike as shown to the right. The ones found on Venus, the Moon and Mars are much, much larger. They go over hills, through riverbeds, through valleys, and over portions of craters. It's almost like they are alive. Sometimes they go up one side of a crater and down another side. Plasma cosmologist Anthony Peratt estimated that a single such bolt to create one of the rilles found on Mars would be as powerful as a 300,000-megaton nuclear explosion. One of the scars found on Mars is shown below attesting to the probability of a war with nasty weapons.

Credibility Note

Certainly, some of the above examples are truly anomalies produced by wind, but the more of these things that pop up, the more it appears to be likely that the wind could not do them all. Some must have been done by humans. I don't mean prehistoric, ape-like beings, I mean space age people. After the massive war and destruction was anyone or anything left? Let's look for vegetation.

Martian Trees

Having a reasonable timeline of some of the events that occurred on Mars suggests that it began losing its atmosphere about the same time as the earth began to spin faster. If that is so, humans stationed on an outpost such as Mars would have tried to conserve the atmosphere by means similar to our terraforming concepts. With the atmosphere mostly carbon dioxide, we can believe the plant was covered with plantlife.

Martian Terra-forming

The humans on Mars may have begun the repair of Mars in our fairly recent past or Mars may be fixing itself. It's been a long time since life could exist on Mars, but there is new evidence that life may be reemerging in the form of some type of plant life. Below is an area many believe is covered with some type of treelike plants. The color and density of these things change over a yearly cycle and they seem to have limbs like trees.

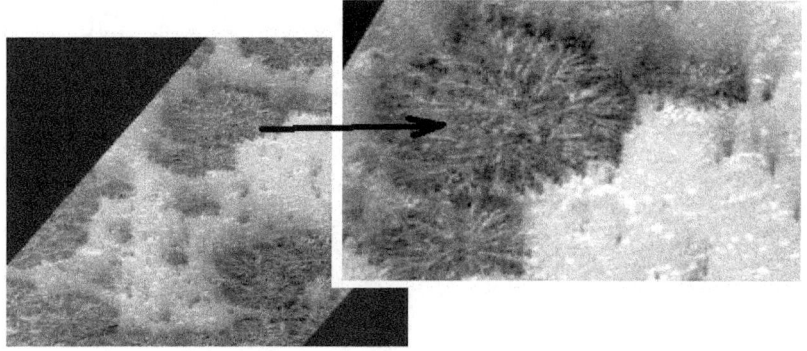

A larger section of this crazy looking area is shown on the following page. There can be little doubt that something is or was living in this area and that it is a huge area.

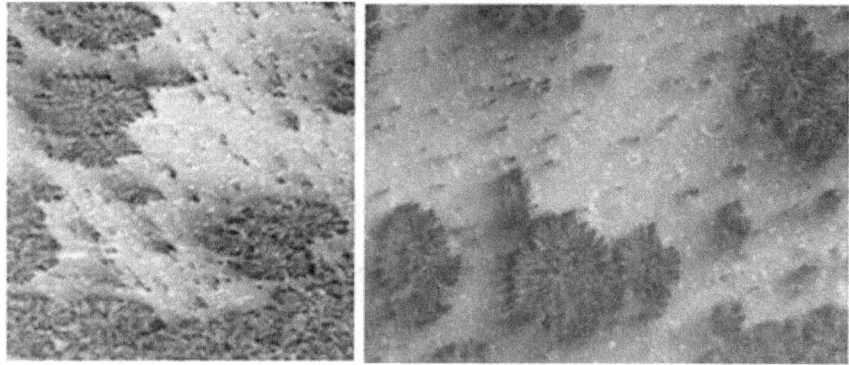

Besides the trees, one area in particular looks to be covered with shrubs.

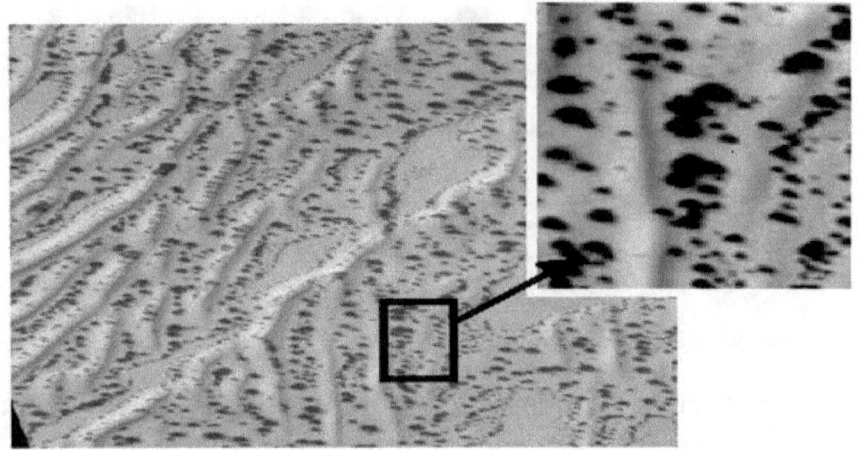

A tiny portion of the once more reasonable oxygenated air the atmosphere, is still surrounding the planet, but now it contains 95% carbon dioxide, so, the plants, at least could breathe and survived to some extent. More possible scrubs dot the landscape. Some envision these objects as rocks, but there is no place for them to have come from. Sprawled out over a large area, these objects dot the landscape with characteristics very similar to desert shrubbery on earth.

As we look at the huge expanse of dune areas, something curious can be found at the base of some of the dunes. Round patches in the Mars photo shown below look very much like some type of vegetation to me.

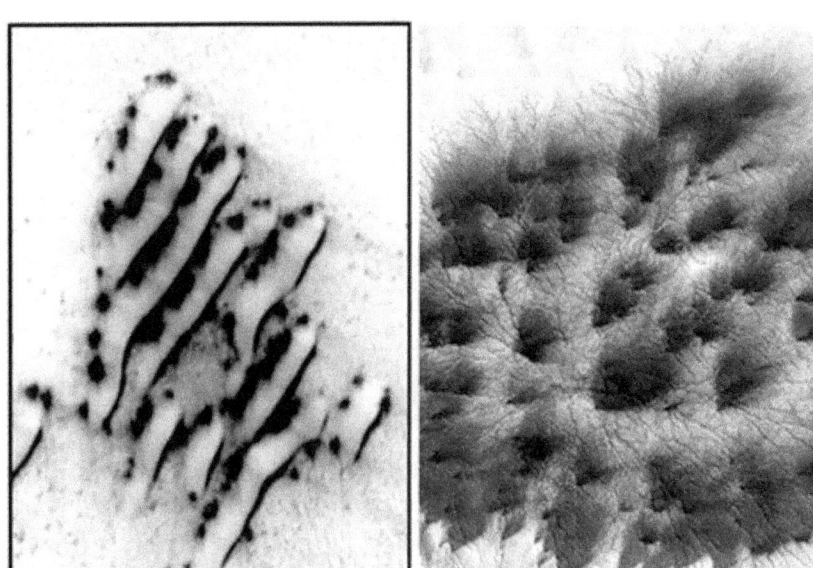

The image to the left above, shows the Martian sagebrush. The dry cracked, often cold surface, somehow holds plants as they cling to life and try to get the limited moisture.

Massive Forest

The next forest shows a dense area with additional tree-like things beginning to grow in front of the original forest. Some say they look more this round signs on sticks, but I still see vegetation.

I blew up some of the ones in front to see if you could figure out what type of plants these are.

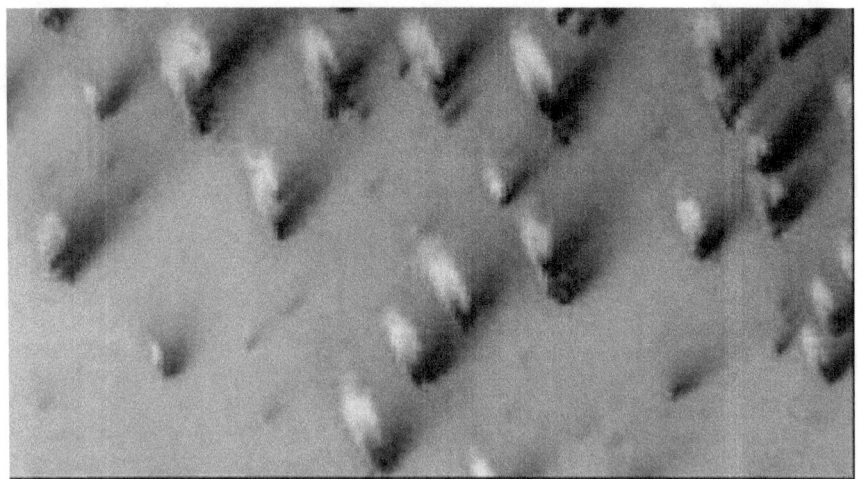

More Trees

The image following is still another clump of what appears to be growing plants of some type. Like the other images, I blew up a section so you could get a better felling of the types of plant-life that is on Mars. Possibly the Martians bring some of these plants underground to help create an oxygen-rich environment below the ground. I added the green color for effect.

Another Forested Area

It certainly appears that whatever is going on it is beginning to work as what appears to be large forested areas are being surrounded by more lightly growing extensions. And the next shows even more vegetation.

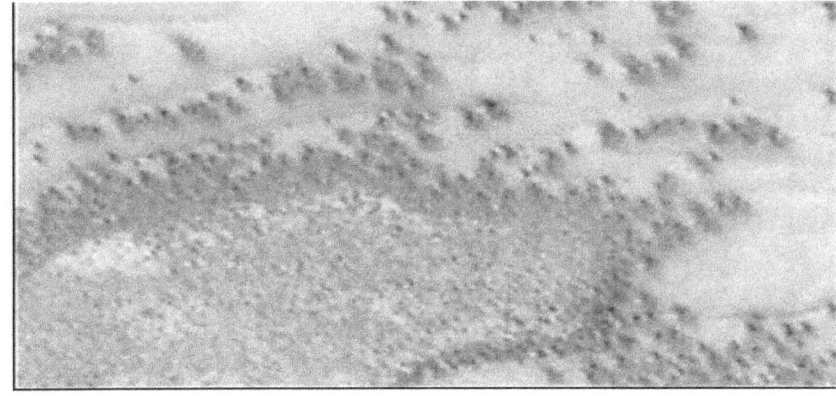

Burned Out Trees

The next images show what appears to be burned out trees all over the place, but possibly it is the only way trees can live on Mars to conserve every drop of water. Possibly these are fed by underground channels to try to regain some semblance of atmosphere over the planet. With regards to trees; look at the dead "looking" trees everywhere.

A slightly closer look at a different spot shows the small clumps of life.

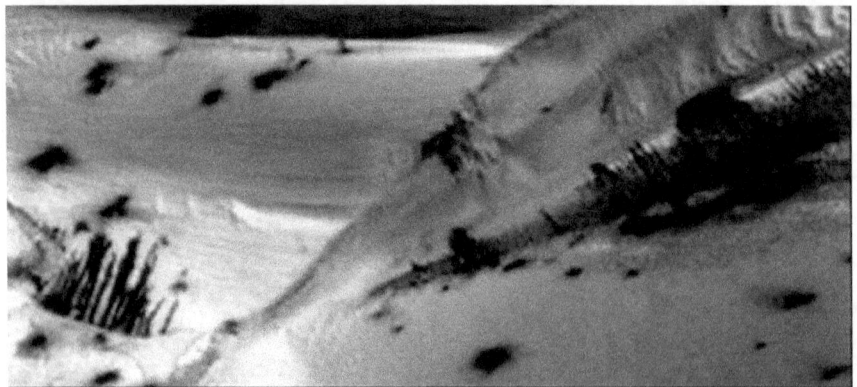

Still another section seems to show these plants form on top of small hill areas and not much life is away from the hills

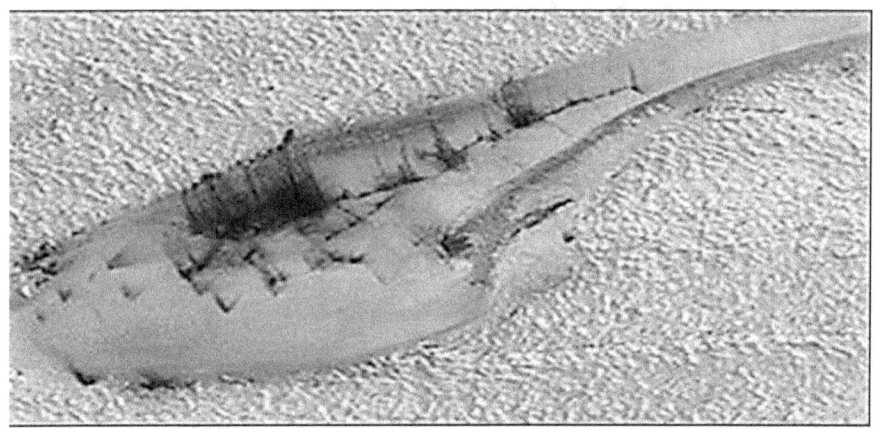

If we blow up a section, the trees can really be examined easily.

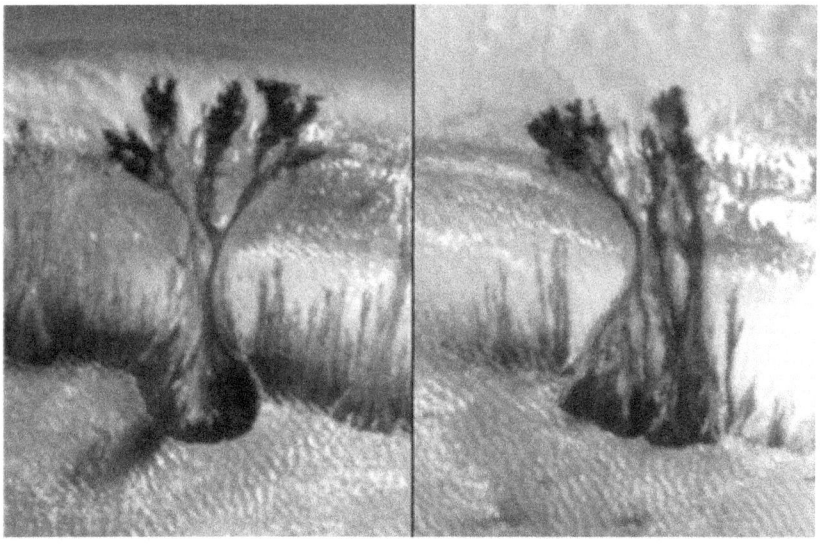

Some places look like the wind has exposed the intricate root systems of many trees and killed them.

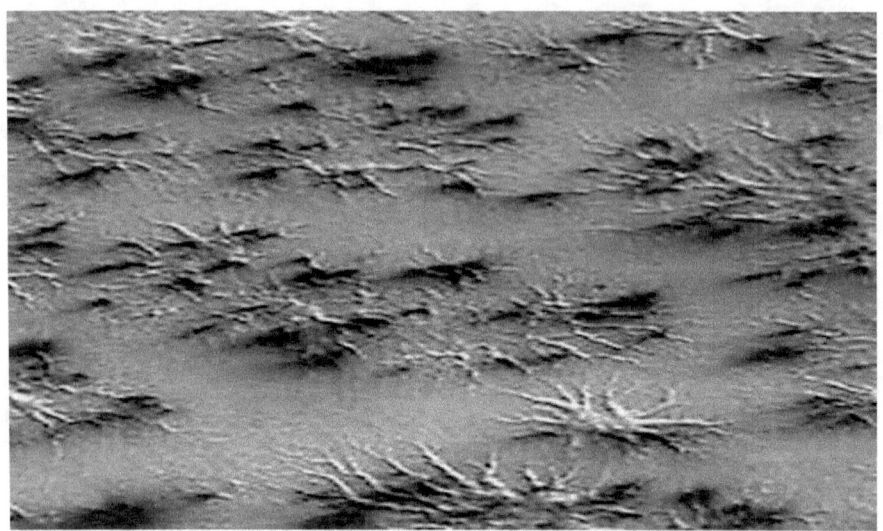

Other areas seem to show vegetation painfully surviving next to what was left of the open sea, no not much more than moist ground.

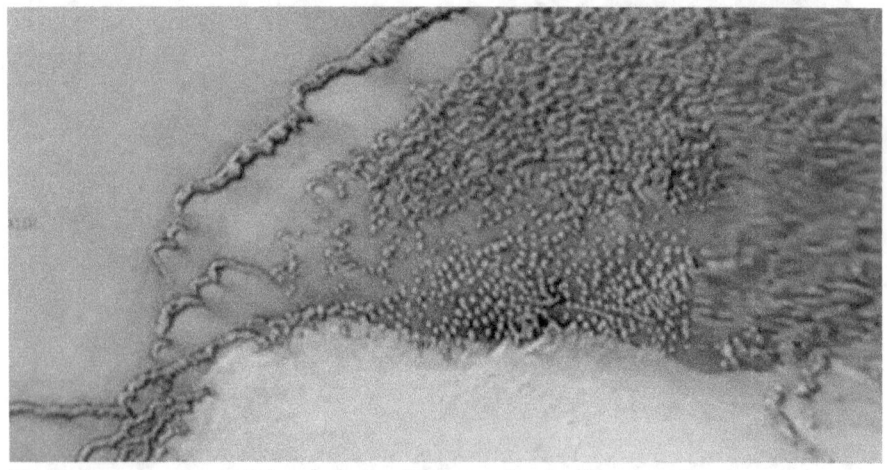

You know the old saying. If you have burned up trees, you can probably find animals so let's see what our rover and orbiting satellites have found.

Herds of Animals

If there are flying ships, roads, vehicle, remains of jets, Forests, water, houses, industrial areas, and tunnels, what about living creatures? The following look like a herd of sheep running along the Martian plains, but some think these are plants rather than animals.

The thing that makes this image interesting is it changes each time it is photographed and so is the following strange image.

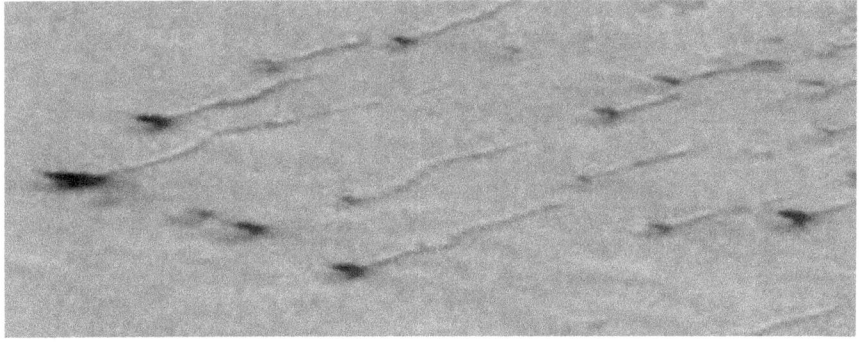

If that is not enough, massive "wormy things" stay near what appears to be a shallow sea.

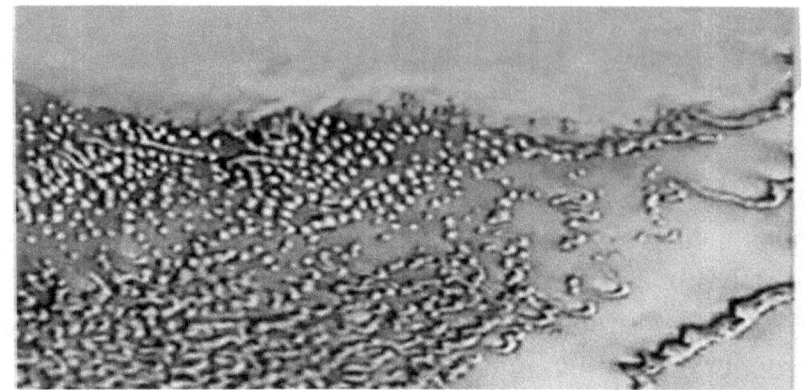

We also must at least consider the large number of images that appear to be human as well. I know I brought this up earlier, but to me, these images look human.

If we go smaller, we find something else. Initially thought to be a spring of some kind, this thing disappeared and now is thought to have been a surfaced worm of sorts.

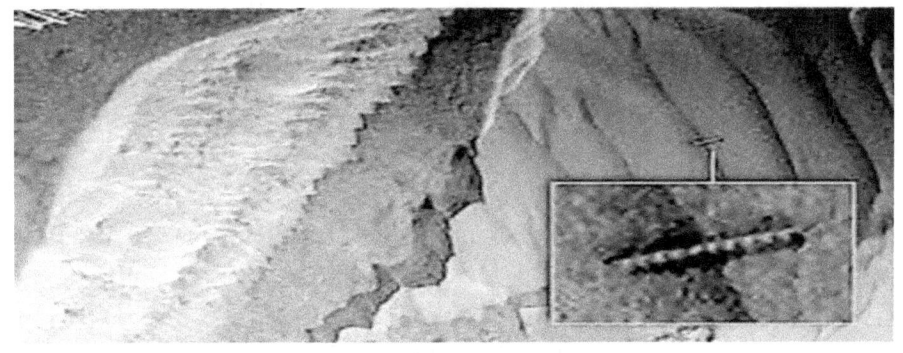

Methane on Mars

What if I told you there is huge amounts of animal produced gases on Mars? Besides all of the previous signs, people still have a hard time with Mars ever being inhabited, but there is methane. We find a significant amount of methane of Mars along the entire equatorial regions. Methane is produced by animals, so animals must have been there. It is in significant amounts so the animals are still around. Possibly tiny micro-organisms, but it does show there is still life there.

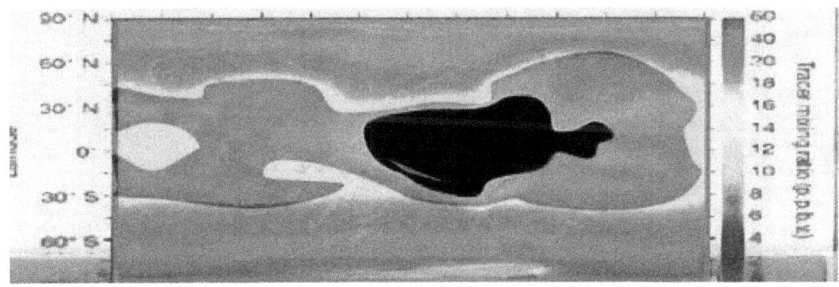

Before we get to more modern interactions with Mars, let's quickly look at the strange moon called Phobos.

Phobos Mysteries

Even the largest Martian moon is strange. On Phobos, unusual features seem to indicate intelligent life. The picture on the following page show some of the odd things which include "square edged" pillars and tracks leading away from its major crater. The tracks are unusual in that they remain equal distance from one another and extend for many miles beyond the crater face. The tracks could possibly be roadways that extend for hundreds of miles. These equal distant tracks have also been found in abundance on Triton, one of the moons of Neptune and on Europa, one of Jupiter's moons.

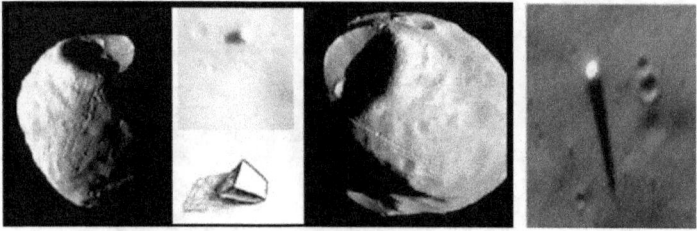

A Taller Pillar- In the same area as the small pillar is a much larger one shown to the left. Note the extremely long shadow. When compared with the small one, it is at least 10 to 12 times as high. Both have squared corners and stick straight up in the air. How could a column extending a mile straight up in the air be natural?

Phobos will soon crash into Mars- According to scientists, within a very short time, the ever-diminishing orbit of Phobos will send it crashing on to the surface. The image below shows the tiny moon next to Mars. Some believe the entire moon to be the remains of a massive man-made orbiter.

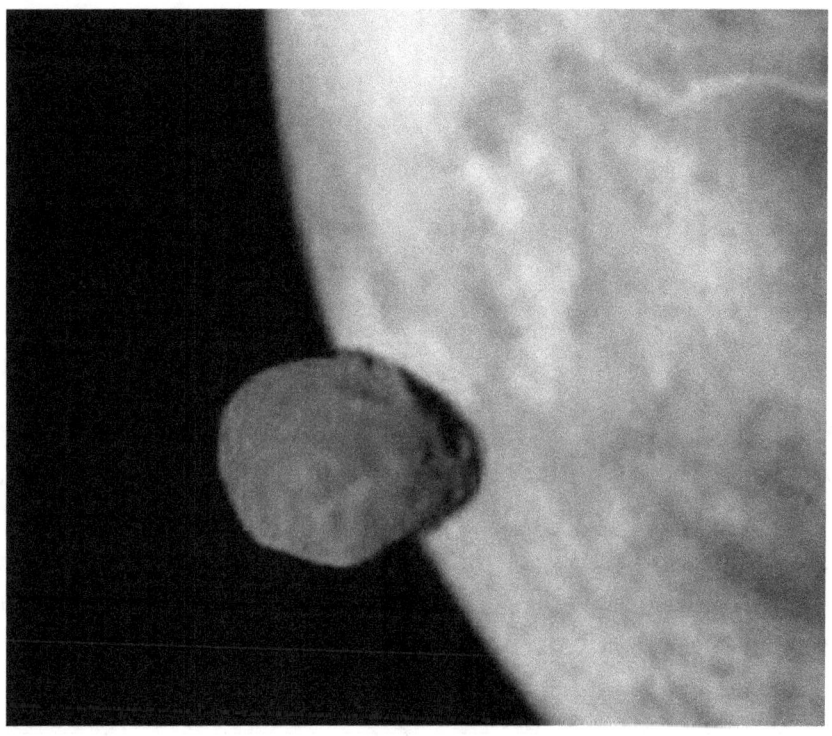

Today, Mars seems to be the most likely place to have once held a human population. Its artifacts seem to be the most geometrical, and certainly the most plentiful. The water that still remains on the planet has been estimated to be only 10 percent of what it was in the not too distant past and with that water there would have been an atmosphere, plants, and almost assuredly it held humans. It looks like much has been destroyed, but when?

When Was It Destroyed?

Brazilian Evidence

As I brought out earlier, the traditions of the ancient Brazilian tribe, Ugha Mongulala, may have given us the insight to when the final war on Mars was significant. We can assume that it occurred either 11 thousand years ago, when the destruction on Venus was so well documented and evidenced, or during the last of the major wars about 6 thousand years ago. Let's review what they told us.

The Golden Age was a time when the gods still ruled over a vast empire on Earth. The moon that we know began to approach the Earth and to circle around it, thousands of years ago, but before the moon took its place the world bore another face. Both lands [Eurasia and the Americas?] were buried under an enormous tidal wave during the first Great Catastrophe [Pleistocene Extinction]. It occurred at the end of the war between the two divine races. The war between the two divine races did not only lay waste to the earth, but also to the worlds of Mars and Venus.

It tells us that both Mars and Venus were laid waste before the great worldwide flood of Noah and the Pleistocene Extinction [10 thousand years ago] and that both planets were affected equally. If we use this data, the Martian cities were destroyed 11 thousand years ago. Venus was completely destroyed, but Mars seemed to have come out of the war in one piece so where is everyone?

Very Little Air

We have to review the history of Mars just a little here. While 400 thousand years ago, the planet was split in half and it began spinning faster just like earth had done after it lost the portion of the planet that was where the Pacific Ocean is today. As I mentioned before, spinning faster does something bad. It makes the planet atmosphere leave. With Mars spinning as fast as Earth does and being only 10% the mass, Mars cannot sustain a breathable atmosphere today. Possibly, it rotated somewhat slower before the massive wars. One possibility is that, while the planet didn't turn into a blazing inferno like Venus, something horrible happened, just the same. We can believe that Mars began rotating faster because of the injected energy into the small planet. This would have slowly driven off the atmosphere and water from the planet until it has about 10% of the water which could be saved underground, but the air wasn't quite as lucky.

The atmosphere of Mars is the layer of gases surrounding Mars composed mostly of carbon dioxide showing the higher concentration of plant life over other biologics. The atmospheric pressure on the Martian surface is about 1% that of Earth so everything would be lighter, for sure. Besides the 95% carbon dioxide [about 100 times what we have on Earth and great for plant-life] in the air there is about, 2% argon, 2% nitrogen, and less than 1 % free oxygen [about 5% as much as here on Earth], 1% carbon monoxide, and 1% methane. The oxygen and methane both suggest a small group of biologics like those on earth. Even those people who were not killed during the final war would soon have had to evacuate the planet before all the water and air were completely gone or retreated underground to manufacture the vital products needed for survival. Even by the war 11 thousand years ago many would have already left. Since that time, the air has been getting even thinner than it was back then.

No matter how and when the air was gone, we know that sometime after the wars damaged large areas of Mars, the people could no longer breathe on the surface. Most left a long time before the air became too thin. If some stayed, they would have had to stay underground. The massive amount of plants could have been used to generate oxygen underneath the surface provided they go get enough water. Perhaps they would have to make sealed tunnels that were somewhat clear to let in the warming rays of the sun. I know it would be almost impossible to find such a thing. This next section may or may not help but I think there are enough witnesses coming forward to, at least let them tell their story. In this case, some people tell us they have been to Mars.

Are There People There Now?

I certainly cannot say definitively there are people on Mars today, but there are stories that we need to, at least, consider. This is a story about something called the Pegasus Project. For this story we must first see what the Web-Bot determined. WebBot is a program that senses probabilities by information scurrying around on the internet. Cliff High is the genius behind this thing and it has been predicting all types of things, mostly in the financial world, but also catastrophic events and other things that would not seem possible to detect by listening to random words and phrases. Anyway! A September 15, 2009 report predicted that a "planetary whistleblower" would emerge from the current period of U.S. financial collapse. It also, somehow, came up with names--- Mr. Basiago, a lawyer from Washington State and Dr. Anderson a Physicist.

Sure enough, there currently is a so-called whistle blower named Andrew Basiago that may tell us a little more about something that sounds bizarre that deals with time travel and Mars. According to testimony, Andrew Basiago, had provided evidence "of a sort" that secret U.S. time travel technologies were used as early as the 1960s. He was recruited when he was a teen to participate in a program called the Pegasus Project that was carried out under a US Defense Advanced Research Projects Agency [DARPA].

During the period 1969 to 1972, he had been describing probes to past and future events that he took via teleportation and something he calls chronovision during the early days of time-space exploration by the US government. Of course, the current Quantum Mechanic Time space jumps confirm the possibility, but the only things we hear about are teleportation of sub-particles.

He indicated that he witnessed the events associate with the destruction of the Twin Towers [September 11, 2001], well before it actually happened. Not only did he view these events, but he has testified that he was sent back to the time of the Gettysburg Address signing and his image was captured in a picture that is still extant. The picture following was supposedly taken November 19 1863. Mr. Basiago was a teen at the time and was suited to be a time traveler. This was just one of the travels he made along with other boys about his same age in the 1970s. We can assume these experiments continued until the present day.

It is not known how the picture came into his possession, but during testimony, Mr. Basiago indicated that he had lost his shoes in transit from something he called the plasma confinement chamber located at East Hanover, NJ in 1972. The destination was Gettysburg, Pennsylvania. Luckily for him, he indicated that a cobbler named John Lawrence Burns saw him and took him into his shop to get a pair of shoes and a Union winter parka to go along with his Bugle boy outfit supplied by DARPA. He indicated that he only was at the location for a short period of time and was returned to New Jersey before he could actually witness any of Lincoln's speech.

I know all this seems bizarre and wacko, but there is more.

Besides this "possible" photograph, Mr. Basiago, shown in the previous picture, also indicated that several people saw his and other boy participants "materialization" at Santa Fe, New Mexico in the 1970s during various test phases of the project. He has indicated that a CIA operative Courtney Hunt worked with him during much of his travels. Her association in a project called Pegasus has been determined, but she has not confirmed any of the details of Mr. Basiago. He also described being teleported from Wood Ridge, NJ to Santa Fe, NM by way of another device. Mr. Basiago said he witnessed an accident in which a boy's feet were sheared off after he was teleported, so there is some danger in these trips.

Dr. Anderson

Like I said, I know everything here sounds hokey, but there is another whistleblower.

His name is Dr. David Lewis Anderson, director of the Anderson Institute. He actually came out before Mr. Basiago and provided a two-hour interview on December 23, 2009 to give an extensive account of his time control research for the U.S. Air Force, which he later continued at his Time Travel Research Institute and other organizations. He is shown middle above. According to this physicist, he was employed at a young age by the U.S. Air Force conducting advanced research and development at the Air Force Flight Test Center at Edwards Air Force Base in the Mojave Desert. During that time, he laid the foundations for what would be known as "time-warp field theory," an approach that was actually used for time travel according to his testimony.

Mr. Cooper

Naval <u>Intelligence</u> officer Milton William Cooper corroborated U.S. Mars bases and time travel accounts of DARPA whistleblower Basiago, Dr. Anderson, and CIA whistleblowers Bernard Mendez and William Stillings. He admitted to key aspects of accounts made by DARPA whistleblower Andrew D. Basiago and the CIA whistleblowers regarding U.S. secret bases on Mars and secret U.S. government time travel capabilities. Milton. Cooper stated that the U.S had first landed on Mars on May 22, 1962 and that, by the time the U.S./NASA public space program landed on the moon in 1969, the U.S. already had a moon base there, since the mid-1950s. Here are some of his own words.

"The first moon landing was May 22, 1962 ... or excuse me, that was the first landing on Mars. I'm sorry, May 22, 1962, was the winged probe that used a hydrozine propeller, flew around approximately three orbits and landed on May 22, 1962, was a joint United States/Russian endeavor. The first time that we landed on the moon was sometime during the ... probably middle 50s, because at the time when President Kennedy stated that he wanted a man to set foot on the moon by the end of the decade we already had a base there."

Mr. Relfe

Michael Relfe is still another "informer". [See below left] A former member of the U.S. armed forces, he was recruited to go to Mars in 1976. He stayed there on a secret base for 20 years before returning by one of those teleporting time traveling things, but here is the weird part. In 1996, he supposedly was sent back to 1976 to finish his military tour. While it may never be known the validity of Mr. Basiago's, Dr. Anderson's, Milton Cooper, Bernard Mendez, William Stillings, or Michael. Relfe's statements, one can be pretty sure that the Rainbow Project did not die as the USS Eldridge disappeared during World War II. Maybe there is more evidence that can be validated better.

Arthur Neumann and Project Camelot

Seems like when one whistleblower comes out the field becomes ablaze with those no longer afraid. As reported on "Nolies Radio", another witness indicated that there is a colony on Mars. His name is Arthur Neumann. [See preceding middle] As the whistleblower who reportedly worked in a program called Project Camelot stated on July 25, 2009, at the European Exopolitics Congress in Spain; there is life on Mars and there are bases on Mars. He went on to say, *"I have been there."* He also provided details of teleporting, a permanent Mars colony, and participating in a one-hour project meeting, which was also attended by representatives of an intelligent civilization that lives in cities under the surface of Mars. This was during a July 26, 2009 Futuretalk interview. As part of his job, Mr. Neumann spoke about the details of his teleportation to Mars during a one-hour project meeting attended by Martian humanoids at the secret underground Mars colony. This not only adds to the growing number of eyewitnesses to some type of teleportation and time travel, but it also brings us to the great granddaughter of President Eisenhower.

Laura Eisenhower

Laura Magdalene Eisenhower [See preceding right] indicated that she was a survivor of an attempted recruitment into the secret Mars colony between 2006 and 2007. All 7 of these witnesses seemed to be talking about the same thing. Some group had developed a way to transport through time and space.

I don't think I could describe anything that sounds more absurd. All I can say is the more you hear seemingly ridiculous things and those things do not contradict other statements, the more the things could be true.

Mr. Basiago stated that in the early 1980's, when they went, the U.S. facilities on Mars were rudimentary and resembled the construction phase of a rural mining project. While there was some infrastructure supporting the jump rooms on Mars, there were no base-like buildings like the U.S. base on Mars first revealed publicly by Command Sgt. Major Robert Dean at the European Exopolitics Summit in Barcelona, Spain in 2009.

On Mars they encountered primitive conditions and some type of dangerous animal that evidently killed a large number of the participants.

In class they were told that *"Of the 97,000 individuals that we have thus far sent to Mars, only 7,000 have survived there after five years."*

While it may never be known the validity of Mr. Basiago's, Dr Anderson's, Mr. Relfe's, or the other whistleblower statements, one can be pretty sure that the Rainbow Project [also known as the Philadelphia Project] did not die as the USS Eldridge disappeared during World War II. It seems the Pegasus Project took over where it ended. Maybe there is more evidence that can be validated better. For a moment, let's just say that these descriptions have a level of truth in them. If so, there must be signs on the surface of Mars. That's where tunnels come in.

Huge Surface Tunnels

If you say it didn't matter that there was water, its simply too cold to live there. The answer might again be just below the surface. What if they typically lived underground and had surface tunnels to allow comfortable transport. OK! Not completely comfortable in the frigid lands. If we look at the full length, we see crisscross avenues and entry into deep caverns. Possible protection from a planetary atmosphere that was getting thinner every year.

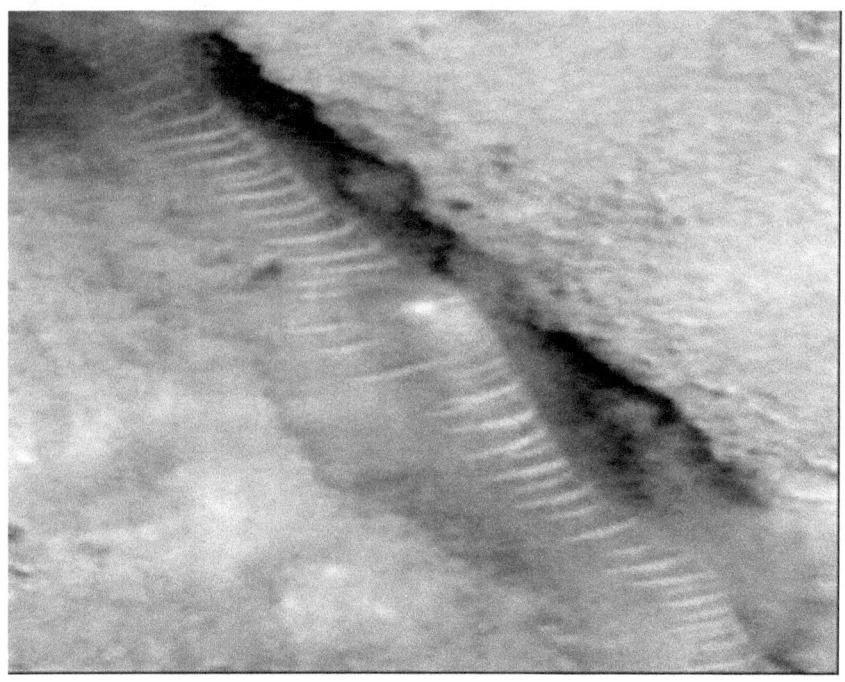

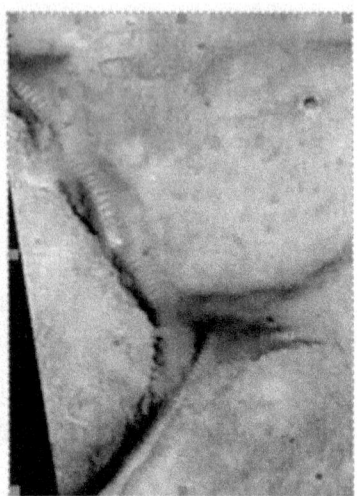

It seems like these tunnels go to some underground location as shown above. Still another section of protective tubes has also been uncovered over time. Where they go, noone knows, but they seem to go somewhere.

If we look at the sections of the preceding tunnels, we see some sections that are constructed in an accordion shape to allow for substantial expansion. All we see is the ribbing on the second one below.

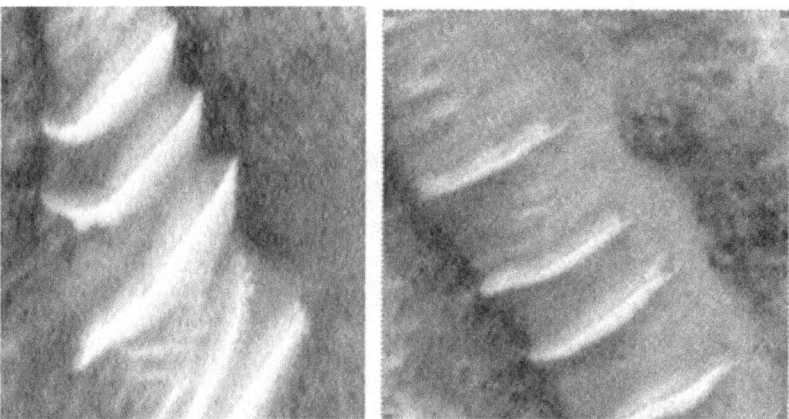

Other sections reveal heavy duty rib designs to ensure maximum structural integrity. Again, it appears we only see the ribbing in this section.

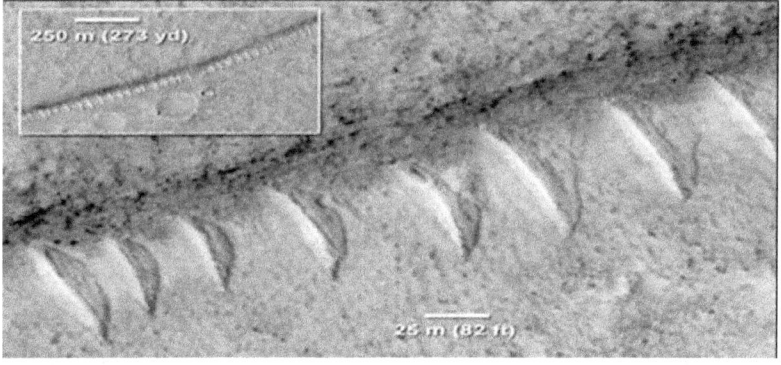

More pictures are below showing how many of these covered tunnels have been found so far.

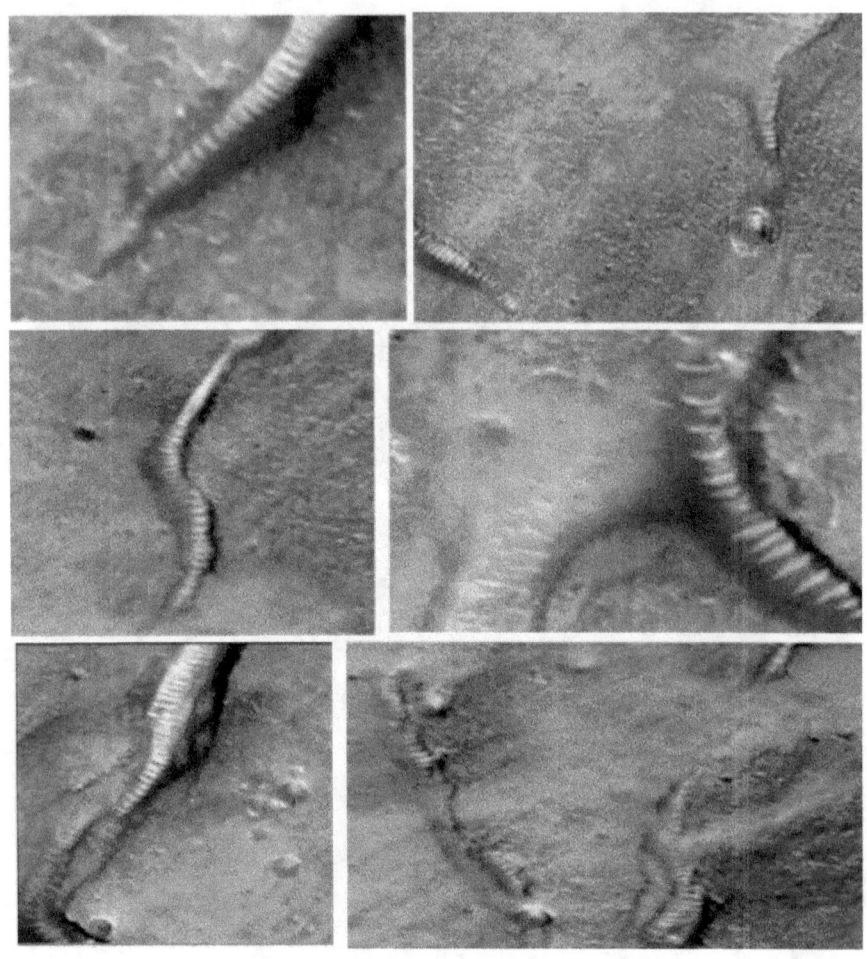

Some Failures

Let's look at the artifact next which is part of the MOC image M1501228. Many believe this to be a huge tunnel that is about 300 meters high and 200 meters wide. It is ribbed for strength like the others, but this one didn't hold up. It was breached a long time ago. Maybe the colonists on Mars were living underground during the initial war years just like the people on earth. With <u>radioactive xenon</u> and <u>melted cities</u>, the <u>tunnels</u> and underground living might have been their only hope. Reflecting light off the surface of the tunnel is evident in the photo which suggests a very smooth surface. No one knows what the tunnel was used for, but not many dispute that this feature would be almost impossible to be produced from natural phenomenon. If this rupture had happened hundreds of thousands of years ago, the cavity would have filled in and there is clearly an open area below the covering, so the tunnel would have been operational during the time of the 1st and second creations of men. Whatever happened to this tunnel was tremendous, just imagine the explosion as a couple thousand meters of structure gave way in a devastating way. Certainly some lost their lives when this important Martian protection was destroyed.

Let's look at some of the details of this marvelous find. Note the clear domed passageway as it can be seen in the fissure.

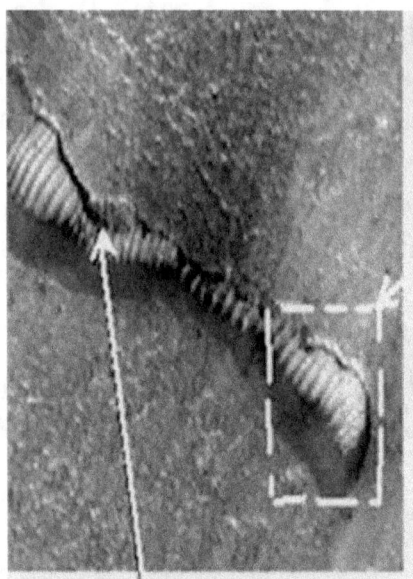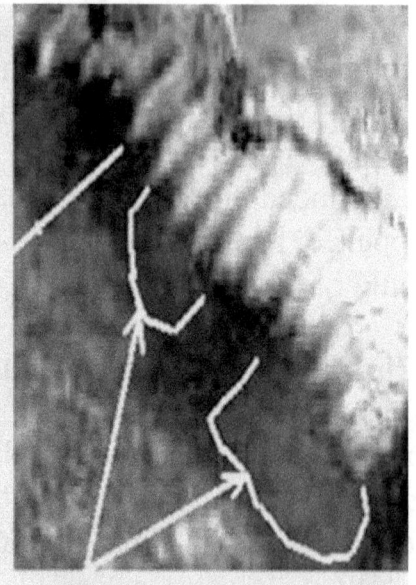

Hollow tube - well defined dark
shadow inside ribbed tube

Reflections into the
shadow

So, what does all this mean??????

Conclusions

I don't know about you, but to me there can be little doubt that people used to live on the planet Mars. These were not aliens or little green men. We are told that normal humans colonized this area and eventually even this outpost was involved in war as power hungry men continuously tried to destroy our own world. While war was a devastating event on Mars, the continuous reduction in atmosphere year after year soon got to be a major concern and potentially the underground tunnels were built as a last-ditch effort to save the colonists.

While evidence indicates that plant life is doing reasonably well in the 95% carbon dioxide world, people have vanished, at least from the surface. With trees all over, there seems to be a reasonable way to manufacture oxygen in the underground homes of the colonist. All but forgotten, new investigations now are uncovering a world lost in time showing us about our ancestors and the battles they faced in colonizing one of our closest planets.

We found massive tunnels, substantial water, plants everywhere, signs of nuclear explosions, and burned out buildings. We have seen flying objects circling the planet, and what appear to be ancient cities being lost in the battle of time and where someone has left a large valve for everyone to see. I know the tennis show and Sasquatch are components of illusion, but all of these features simply cannot be.

The ancient texts tell about the colonies on mars being destroyed when Venus met its doom 11 thousand years ago, so we can believe some level of existence until that time as the ancient text data is verified by extremely old artifacts showing flying machine space outfits.

Then there is project Pegasus. While is sounds bizarre, it seems to go along with the Rainbow project where the USS Eldridge disappeared from a Philadelphia harbor after Tesla walked away from what he considered too dangerous. Possibly, the disappearance of the USS Eldridge was just the beginning and we now have working laboratories on the surface.

Some of the people on the planet may have big eyes, a tiny nose, and small features from years of breathing high oxygen content air in an underground lifestyle on a planet with low gravity, but many of the Martians seem to be just regular people. As a group of colonizers are readying themselves to inhabit a new world, they may find friendly faces.

With that I must end this book.

About The Author

Steve Preston is a long lime author of scientific, esoteric facts. His series on the creation of mankind is shown below. The series focuses on the painful truths rather than whitewashed details that make us comfortable. If you are interested in the truth instead of comfort, please continue to read and, while you are at it, review other works by Mr. Preston as shown below.

Books on War
Four Armageddons
World War Before
World War Zero
World War with Heaven
America's Civil War Lie
Astronomy & Earth Science
Complex Earth
Retiming the Earth
Evolution Of The Planets
Physics
Vibrational Matter
Our 12 Dimensional Universe
Meaning of Light and Life
Walk Through a Wall & Time
Ancient History of Flying
Mystery of Photons and Light
Ancient History
Behind the Tower of Babel
Who Really Discovered the Americas?
When Giants ruled the Earth
Kingdoms Before the Flood
Why Men Lived Underground
Anakim Gods
Why Rome Fought the Berserkers
Development of Mankind Series
The First Creation of Man
The Second Creation of Man
The Creation Of Adam And Eve.
The Antediluvian War Years
Man After the Flood

A Closer Look At Ancient History
A New View Of Modern History
The 20th Century To The End Of Time
Egypt and DNA
Scythians Conquer Ireland
Alien DNA Inside Us
Moses Saved Egypt
Mysteries of the Exodus
Races of Men
Religion
Closer Look At Genesis
Why the King James Bible Failed
New Look At The Bible
History Confirmed By The Bible
Genesis Companion
Adam's First Wife
Sex Crazed Angels
Anthropic Science
Life Resonance
Anthropic Reality
Awaken the Departed
Self, Soul, and Spirit
Contemporary Issues
American School Disaster
Allah' the Moon God
Fast History of MILES Training
Our Very Odd Presidents
Bad Side of Lincoln
Disgusting Display

www.ingramcontent.com/pod-product-compliance
Lightning Source LLC
Chambersburg PA
CBHW070228190526
45169CB00001B/121